A First Course
In Analysis

A First Course
In Analysis

Donald Yau

The Ohio State University at Newark, USA

World Scientific

NEW JERSEY · LONDON · SINGAPORE · BEIJING · SHANGHAI · HONG KONG · TAIPEI · CHENNAI

Published by

World Scientific Publishing Co. Pte. Ltd.

5 Toh Tuck Link, Singapore 596224

USA office: 27 Warren Street, Suite 401-402, Hackensack, NJ 07601

UK office: 57 Shelton Street, Covent Garden, London WC2H 9HE

British Library Cataloguing-in-Publication Data
A catalogue record for this book is available from the British Library.

A FIRST COURSE IN ANALYSIS

ISBN 978-981-4417-85-3

Printed in Singapore.

To Eun Soo and Hye-Min

Preface

This is an introductory text on real analysis for undergraduate students. The first course in real analysis is full of challenges, both for the instructors and the students. Many mathematics majors consider real analysis a difficult course. The transition from mechanical computation to formal, rigorous proofs is difficult even for many mathematics majors. Most students beginning a course in real analysis have never been asked to understand and construct proofs before. Moreover, even if one has some ideas about how a proof should go, writing it down in a logical manner is a challenge in itself. This book is written with these challenges in mind.

The prerequisite for this book is a solid background in freshman calculus in one variable. The intended audience of this book includes undergraduate mathematics majors and students from other disciplines whose use real analysis. Since this book is aimed at students who do not have much prior experience with proofs, the pace is slower in earlier chapters than in later chapters.

In most instances, motivations for new concepts are explained before the actual definitions. For many concepts that have negations (for example, convergence of a sequence), such negations are also stated explicitly. Wherever appropriate we discuss the basic ideas that lead to a proof before the actual proof is given. Such discussion is intended to help students develop an intuition as to how proofs are constructed. There are exercises at the end of each section and of each chapter. Occasionally, some further topics are explored in these additional exercises.

To the Students

There are a few things that you should keep in mind as you work through this book. A professor of mine, who I shall not name here, once told me this: *Nobody teaches you mathematics. You teach yourself mathematics.* You cannot hope to master the materials in this book simply by watching your instructor lecturing. In fact, if you could do that, this book is not for you. I have tried to make the materials as accessible as possible, but you have to do most of the work. You should attempt as many exercises as possible, whether they are assigned homework or not. Expect to do lots of scratch work as you attempt the exercises. After you have written down a solution to an exercise, read it again and again, and then some more, to see if every step is logical.

Expect to think deep and hard as you go through this book. When you read a proof or an example, make sure you understand where each hypothesis is used. Make sure that you understand every single step in a proof. If you get stuck at a certain proof or step, let some time elapse and go back to it later. In particular, do not expect to understand everything the first time you read it. There are many parts in this book that you should read and think through multiple times if you want to master them.

Besides this book, the book [Gelbaum and Olmsted (1964)] is highly recommended as a source of many good and exotic examples.

Donald Yau

Contents

Chapter 1

Sets, Functions, and Real Numbers

There are two main purposes of this chapter. First, in sections 1.1 and 1.2 we fix some notations and introduce some terminology about sets and functions that will be used in the rest of this book. We also discuss the very useful tool of mathematical induction in section 1.4.

Second, basic properties of the real number system are discussed. In section 1.3 the set of real numbers as an ordered field is discussed. All of the ordered field properties of real numbers should already be familiar to the reader. Also, the Completeness Axiom and the Archimedean Property are discussed. In section 1.5 we discuss countable and uncountable sets, both of which are important concepts for real numbers.

It is possible, in fact, logically correct, to construct the real number system with all of its well known properties, starting with some set theory axioms. However, we will not take this path in this book, so most of the basic algebraic properties of the real number system are assumed known. The reader who is interested in a rigorous construction of the real number system may consult [Hobson (1907)], [Rudin (1976)], or [Sprecher (1987)].

1.1   Sets

The purpose of this section is to establish some notations and terminology regarding sets. These concepts will be used throughout the rest of this book. More set theory concepts will be introduced in Section 1.5. A good reference for basic set theory is [Halmos (1974)].

1.1.1   *Set Notations*

By a **set** S we mean a possibly empty collection of objects, called **elements**. If x is an element in S and y is not an element in S, we write $x \in S$ and $y \notin S$, respectively. The set with zero element is called the **empty set** and is denoted by $\varnothing$. A set with at least one element is said to be **non-empty**. We sometimes write $\{x : x \in S\}$ to specific the elements in a set S.

Example 1.1. Here are some examples of sets:

- The set $\mathbb{Z}_+ = \{1, 2, 3, \ldots\}$ of **positive integers**.
- The set $\mathbb{N} = \{0, 1, 2, \ldots\}$ of **natural numbers**.
- The set $\mathbb{Z} = \{0, \pm 1, \pm 2, \ldots\}$ of **integers**.
- The set $\mathbb{Q} = \{\frac{m}{n} : m, n \in \mathbb{Z}, n \neq 0\}$ of **rational numbers**.
- The set $\mathbb{R}$ of **real numbers**.
- The set $\mathbb{I}$ of **irrational numbers**. A real number x is said to be irrational if it is not a rational number.

1.1.2   *Subsets*

Suppose that S and T are sets. If every element in S is also an element in T, then we say that S is a **subset** of T and write $S \subseteq T$, or sometimes $T \supseteq S$. So S is *not* a subset of T if and only if there exists an element $x \in S$ that does not belong to T. The phrase *if and only if* means implications in both directions. So if A and B are two statements, then

$$A \text{ if and only if } B$$

means both A implies B (i.e., if A is true, then B is true) *and* B implies A (i.e., if B is true, then A is true).

   If S is a subset of T and if there exists an element $x \in T$ that does not belong to S, then we write $S \subsetneq T$, or sometimes $T \supsetneq S$, and call S a **proper subset** of T. If S and T have exactly the same elements, then we write $S = T$ and say that they are equal.

Example 1.2. We have the following proper subset inclusions:

$$\mathbb{Z}_+ \subsetneq \mathbb{N} \subsetneq \mathbb{Z} \subsetneq \mathbb{Q} \subsetneq \mathbb{R} \supsetneq \mathbb{I}.$$

The only inclusion that is not obvious is the last one. In other words, is there really a real number x that is not a rational number? The answer is yes, as we will see in section 1.3. One example of an irrational number is $\sqrt{2}$ (Theorem 1.4).

1.1.3   *Operations on Sets*

Let S and T be two sets. There are several ways to build new sets from these two given sets. The **union** is defined as

$$S \cup T = \{x : x \in S \text{ or } x \in T\},$$

an element of which is an element in S or an element in T. For example, we have

$$\{0, 2, 5\} \cup \{1, 2, 4\} = \{0, 1, 2, 4, 5\}.$$

The union of a collection of sets S_n, where $n \in \mathbb{N}$, is defined similarly as

$$\bigcup_{n=1}^{\infty} S_n = \{x : x \in S_n \text{ for some } n \in \mathbb{N}\}.$$

The **intersection** of S and T is defined as

$$S \cap T = \{x : x \in S \text{ and } x \in T\}.$$

So an element in the intersection $S \cap T$ is an element that lies in both S and T. For example, we have

$$\{0, 2, 5\} \cap \{1, 2, 4\} = \{2\},$$

which consists of the single element 2. The intersection of a collection of sets S_n, where $n \in \mathbb{N}$, is defined similarly as

$$\bigcap_{n=1}^{\infty} S_n = \{x : x \in S_n \text{ for every } n \in \mathbb{N}\}.$$

Two sets S and T are said to be **disjoint** if their intersection $S \cap T$ is the empty set. If S and T are disjoint, we call $S \cup T$ their **disjoint union**.

The **difference** is defined as

$$S \smallsetminus T = \{x : x \in S \text{ and } x \notin T\}.$$

It consists of the elements in S that are not in T. For example, we have

$$\{0, 2, 5\} \smallsetminus \{1, 2, 4\} = \{0, 5\}.$$

The **Cartesian product** is defined as

$$S \times T = \{(x, y) : x \in S \text{ and } y \in T\}.$$

It consists of the ordered pairs whose first entry lies in S and whose second entry lies in T. For example, we have

$$\{a, b\} \times \{c, d\} = \{(a, c), (a, d), (b, c), (b, d)\}.$$

The Cartesian product of a collection of sets S_n, where $n \in \mathbb{N}$, is defined similarly as

$$\prod_{n=1}^{\infty} S_n = \{(x_1, x_2, \ldots) : x_n \in S_n \text{ for each } n \in \mathbb{N}\}.$$

1.1.4 *Exercises*

In the exercises below, the symbols A, B, C, S, T, etc., denote arbitrary sets.

(1) Are the sets $\varnothing$ and $\{\varnothing\}$ equal? Justify your answer.
(2) Prove that $S = T$ if and only if both $S \subseteq T$ and $T \subseteq S$.
(3) Prove that $S \subseteq T$ if and only if $S \cap T = S$.
(4) Prove the *Distributive Laws*:

   (a) $A \cup (B \cap C) = (A \cup B) \cap (A \cup C)$.

(b) $A \cap (B \cup C) = (A \cap B) \cup (A \cap C)$.

(5) Prove the following generalizations of the Distributive Laws:

   (a) $A \cup (\bigcap_{i=1}^{\infty} B_i) = \bigcap_{i=1}^{\infty} (A \cup B_i)$.
   (b) $A \cap (\bigcup_{i=1}^{\infty} B_i) = \bigcup_{i=1}^{\infty} (A \cap B_i)$.

(6) Prove the *DeMorgan Laws*:

   (a) $S \setminus (A \cup B) = (S \setminus A) \cap (S \setminus B)$.
   (b) $S \setminus (A \cap B) = (S \setminus A) \cup (S \setminus B)$.

(7) Prove the following generalizations of the DeMorgan Laws:

   (a) $S \setminus (\bigcup_{i=1}^{\infty} A_i) = \bigcap_{i=1}^{\infty} (S \setminus A_i)$.
   (b) $S \setminus (\bigcap_{i=1}^{\infty} A_i) = \bigcup_{i=1}^{\infty} (S \setminus A_i)$.

(8) Prove that $S \times (\bigcup_{n=1}^{\infty} A_n) = \bigcup_{n=1}^{\infty} (S \times A_n)$.

(9) Prove that $(A \times B) \cap (S \times T) = (A \cap S) \times (B \cap T)$.

(10) For a set S, define its **power set** $\mathcal{P}(S)$ to be the set whose elements are the subsets of S. For example,

$$\mathcal{P}(\{a,b\}) = \{\varnothing, \{a\}, \{b\}, \{a,b\}\}.$$

If S has n elements, where n is an arbitrary natural number, find the number of elements in its power set $\mathcal{P}(S)$.

1.2   Functions

The purpose of this section is to establish some notations and terminology regarding functions.

1.2.1   *Functional Notations*

The concept of a function should already be familiar from calculus. Let S and T be two sets. A **function** from S to T, written as

$$f : S \to T,$$

is a rule that assigns to each element $x \in S$ an element $f(x) \in T$. Equivalently, such a function f is a subset U of the Cartesian product $S \times T$ such that for every element x in S, there exists exactly one element in U of the form (x,y) for some $y \in T$. The set S is called the **domain** of f, which is denoted by $Dom(f)$. The set T is called the **target** of f. The **range** of f is the subset

$$Ran(f) = \{f(x) \in T : x \in S\} \subseteq T.$$

For a subset A of S, the **image** of A under f is the subset

$$f(A) = \{f(a) : a \in A\} \subseteq T.$$

For a subset B of T, the **inverse image** of B under f is the subset

$$f^{-1}(B) = \{x \in S : f(x) \in B\} \subseteq S.$$

Example 1.3.

(1) There is a function $f: \mathbb{R} \to \mathbb{R}$ defined by $f(x) = 2x - 1$ for real numbers x. The range of f is all of $\mathbb{R}$. The image of $\mathbb{Z}_+$ under f is

$$f(\mathbb{Z}_+) = \{1, 3, 5, \ldots\}.$$

The inverse image of $\mathbb{Z}_+$ under f is

$$f^{-1}(\mathbb{Z}_+) = \left\{1, \frac{3}{2}, 2, \frac{5}{2}, \ldots\right\} = \left\{1 + \frac{n}{2} : n \in \mathbb{N}\right\}.$$

(2) There is a function $g: \{0, 2, 5\} \to \{1, 2, 4\}$ defined by $g(0) = g(2) = 4$ and $g(5) = 1$. The range of g is the proper subset $\{1, 4\}$ of $\{1, 2, 4\}$. Moreover, we have $f^{-1}(\{2\}) = \varnothing$ and $f^{-1}(\{4\}) = \{0, 2\}$.

1.2.2 Special Functions

Definition 1.1. Let $f: S \to T$ be a function.

- We call f an **injection** if for elements x and y in S,

$$f(x) = f(y) \quad \text{implies} \quad x = y.$$

In this case, we also say that f is **injective**.
- We call f a **surjection** if for every element $z \in T$, there exists an element $x \in S$ such that $f(x) = z$. In this case, we also say that f is **surjective**.
- A function that is both an injection and a surjection is called a **bijection**.

Thus, a function $f: S \to T$ is *not* injective if and only if there exist two distinct elements x and y in S such that $f(x) = f(y)$. A function $f: S \to T$ is *not* surjective if and only if there exists an element $z \in T$ that does not lie in the range of f.

If $f: S \to T$ is a bijection, then there is a bijection $g: T \to S$ (Exercise (13)). In other words, there exists a bijection from S to T if and only if there exists a bijection from T to S.

A function can be injective without being surjective, or surjective without being injective. Also, it can be neither injective nor surjective. These statements can be demonstrated with the following examples.

Example 1.4. Suppose $f: \mathbb{R} \to \mathbb{R}$ is a function.

(1) If $f(x) = 2^x$, then it is injective but not surjective.
(2) If $f(x) = x(x-1)(x+1)$, then it is surjective but not injective.
(3) If $f(x) = x^2$, then it is neither injective nor surjective.
(4) If $f(x) = 2x - 1$, then it is a bijection.

1.2.3   *Inverse Functions*

Suppose that $f: S \to T$ is an injection. In other words, if x and y are two distinct elements in S, then $f(x) \neq f(y)$ in T. In this case, there is an **inverse function**

$$f^{-1}: Ran(f) \to Dom(f)$$

defined by

$$f^{-1}(z) = x \quad \text{if and only if} \quad f(x) = z \tag{1.1}$$

for all $x \in S = Dom(f)$ and $z \in Ran(f)$.

Example 1.5. The function $f: \mathbb{R} \to \mathbb{R}$ defined as $f(x) = 2x - 1$ is a bijection, which is, in particular, an injection. Its inverse function $f^{-1}: \mathbb{R} \to \mathbb{R}$ is given by $f^{-1}(x) = \frac{1}{2}(x + 1)$ for any real number x.

If f is a bijection, then its inverse function f^{-1} is also a bijection. Moreover, the inverse function $(f^{-1})^{-1}$ of f^{-1} is f. This is Exercise (13) below.

1.2.4   *Composition*

If $f: S \to T$ and $g: T \to U$ are functions, then their **composition**

$$g \circ f: S \to U$$

is defined as

$$(g \circ f)(x) = g(f(x))$$

for elements x in S. Note that in order to define the composition $g \circ f$, the range of f must be a subset of the domain of g.

Example 1.6. Consider the functions $f, g: \mathbb{R} \to \mathbb{R}$ defined by $f(x) = x^2$ and $g(x) = 2x - 1$. In this case, we can form the compositions

$$(g \circ f)(x) = g(x^2) = 2x^2 - 1 \quad \text{and} \quad (f \circ g)(x) = (g(x))^2 = (2x - 1)^2.$$

1.2.5   *Exercises*

(1) If $f, g: S \to S$ are two functions such that $f \circ g = g \circ f$, does it follow that $f = g$?

(2) Let $f: S \to T$ and $g: T \to U$ be functions.

    (a) If both f and g are injections, prove that $g \circ f$ is also an injection.

    (b) If both f and g are surjections, prove that $g \circ f$ is also a surjection.

    (c) If both f and g are bijections, prove that $g \circ f$ is also a bijection.

(3) Let $f: S \to T$ and $g: T \to U$ be functions.

    (a) If $g \circ f$ is an injection, prove that f is an injection.

    (b) If $g \circ f$ is an injection, does it follow that g is an injection.

(4) Let $f: S \to T$ and $g: T \to U$ be functions.

(a) If $g \circ f$ is a surjection, prove that g is a surjection.

(b) Give an example in which $g \circ f$ is a surjection, but f is not a surjection.

(5) For a positive integer n, determine the number of bijections from the set $\{1, \ldots, n\}$ to itself.

(6) Let m and n be positive integers.

(a) How many different functions $f: \{1, \ldots, n\} \to \{1, \ldots, m\}$ are there?

(b) Suppose that $n \leq m$. How many functions f in the previous part are injections?

(7) Let $f: S \to T$ be a function, and let A and B be subsets of S.

(a) Prove that $f(A \cup B) = f(A) \cup f(B)$.

(b) Prove that $f(A \cap B) \subseteq f(A) \cap f(B)$.

(c) Prove that, if f is an injection, then $f(A \cap B) = f(A) \cap f(B)$.

(d) Is the equality $f(A \cap B) = f(A) \cap f(B)$ always true?

(8) Let $f: S \to T$ be a function, and let A and B be subsets of T.

(a) Prove that $f^{-1}(A \cup B) = f^{-1}(A) \cup f^{-1}(B)$.

(b) Prove that $f^{-1}(A \cap B) = f^{-1}(A) \cap f^{-1}(B)$.

(c) Prove that $f^{-1}(T \smallsetminus A) = S \smallsetminus f^{-1}(A)$.

(9) Let $f: S \to T$ be a function, and let $A_1, A_2, \ldots$ be subsets of T.

(a) Prove that $f^{-1}\left(\bigcup_{n=1}^{\infty} A_n\right) = \bigcup_{n=1}^{\infty} f^{-1}(A_n)$.

(b) Prove that $f^{-1}\left(\bigcap_{n=1}^{\infty} A_n\right) = \bigcap_{n=1}^{\infty} f^{-1}(A_n)$.

(10) Let $f: S \to T$ be an injection. Prove that it has a unique inverse function.

(11) In each case, (i) verify that f is an injection, (ii) find the inverse function f^{-1} of f, and (iii) specify domain and range of f^{-1}.

(a) $f(x) = \sqrt{x}$ for $x \geq 0$.

(b) $f(x) = \sqrt{2 + 3x}$ for $x \geq -2/3$.

(c) $f(x) = 1 + x^2$ for $x \geq 0$.

(d) $f(x) = 2x + 5$ for $x \in \mathbb{R}$.

(12) Let $f: S \to T$ and $g: T \to U$ be injections. Prove that $(g \circ f)^{-1} = f^{-1} \circ g^{-1}$.

(13) Let $f: S \to T$ be a bijection.

(a) Prove that its inverse function f^{-1} is also a bijection.

(b) Prove that the inverse function $(f^{-1})^{-1}$ of f^{-1} is f itself.

(c) Prove that $f^{-1}(f(x)) = x$ and $f(f^{-1}(y)) = y$ for all $x \in S$ and $y \in T$.

(14) Let $f: S \to T$ and $g: T \to U$ be two functions. Suppose that $(g \circ f)(x) = x$ and $(f \circ g)(y) = y$ for all $x \in S$ and $y \in T$.

(a) Prove that f and g are both bijections.

(b) Prove that $g = f^{-1}$.

1.3   Real Numbers

In the section, some basic algebraic properties of real numbers are discussed. Then we state the Completeness Axiom and discuss some of its consequences, including the Archimedean Property and characterization of intervals.

1.3.1   *Ordered Field*

Let x, y, and z be three arbitrary real numbers. There are two basic operations of real numbers. Namely, one can form the sum $x + y$ and the product xy. The following properties should be familiar to the reader.

(F1) $x + y = y + x$.
(F2) $(x + y) + z = x + (y + z)$.
(F3) There exists an element 0 such that $x + 0 = x$.
(F4) There exists a real number $-x$ such that $x + (-x) = 0$.
(F5) $xy = yx$.
(F6) $(xy)z = x(yz)$.
(F7) There exists an element 1 such that $1 \cdot x = x \cdot 1$.
(F8) If $x \neq 0$, then there exists an element x^{-1} such that $x \cdot x^{-1} = 1$.
(F9) $x(y + z) = xy + xz$.

The first two properties state that addition is commutative and associative. The properties **(F3)-(F4)** state that there exists a 0 real number, and every real number has an additive inverse. The properties **(F5)-(F8)** are the corresponding statements for multiplication. The last property says that multiplication is distributive over addition.

A set S equipped with two operations + (addition) and × (multiplication) satisfying the properties **(F1)-(F9)** is called a **field**. So the set $\mathbb{R}$ of real numbers is a field, as is the set $\mathbb{Q}$ of rational numbers (Exercise (1)).

Given two arbitrary real numbers x and y, it is possible to compare them using the inequality $\leq$ (less than or equal to). The following properties of this relation should be familiar to the reader. Again, x, y, and z are arbitrary elements in $\mathbb{R}$.

(O1) Either $x \leq y$ or $y \leq x$; both of these happen at the same time if and only if
$x = y$.
(O2) If $x \leq y$ and $y \leq z$, then $x \leq z$.
(O3) If $x \leq y$, then $x + z \leq y + z$.
(O4) If $x \leq y$ and $0 \leq z$, then $xz \leq yz$.

If $x \leq y$ and $x \neq y$, then we write $x < y$ and say that x is strictly less than y. We will also use the notations $y \geq x$ for $x \leq y$ and $y > x$ for $x < y$. For real numbers $x < 0 < y$, we say that x is **negative** and y is **positive**.

A field F equipped with a relation $\leq$ satisfying the properties **(O1)-(O4)** is called an **ordered field**. So the set $\mathbb{R}$ of real numbers is an ordered field, as is the set $\mathbb{Q}$ of rational numbers.

Given a finite number of real numbers $\{x_1, \ldots, x_n\}$, their **maximum** and **minimum** elements are denoted by $\max\{x_1, \ldots, x_n\}$ and $\min\{x_1, \ldots, x_n\}$, respectively.

1.3.2   *Absolute Value*

Another familiar property of real numbers is the absolute value.

Definition 1.2. Let x be a real number. Its **absolute value** is defined as the real number

$$|x| = \begin{cases} x & \text{if } x \geq 0, \\ -x & \text{if } x < 0. \end{cases}$$

Note that

$$|x| \leq y \quad \text{if and only if} \quad -y \leq x \leq y. \tag{1.2}$$

In particular, applying this to $y = |x|$, we have

$$-|x| \leq x \leq |x| \tag{1.3}$$

for any real number x.

The most important property of the absolute value that we will use is the Triangle Inequality. Recall from calculus that if $\mathbf{a}$ and $\mathbf{b}$ are two vectors, one can form a triangle in which the edges are $\mathbf{a}$, $\mathbf{b}$, and $\mathbf{a}+\mathbf{b}$. The sum of the lengths of the first two edges, $|\mathbf{a}| + |\mathbf{b}|$, is at least the length of the remaining edge $|\mathbf{a}+\mathbf{b}|$. The Triangle Inequality is the real number analog of this fact.

Theorem 1.1 (Triangle Inequality). *For any real numbers x and y, we have*

$$|x + y| \leq |x| + |y|.$$

Proof. Adding the inequalities

$$-|x| \leq x \leq |x| \quad \text{and} \quad -|y| \leq y \leq |y|,$$

we obtain

$$-(|x| + |y|) = -|x| - |y| \leq x + y \leq |x| + |y|.$$

Using the characterization (1.2), the above inequalities are equivalent to the Triangle Inequality. $\square$

1.3.3 *Upper and Lower Bounds*

The Completeness Axiom of $\mathbb{R}$ is about the existence of a certain upper bound of sets of real numbers. Before discussing this axiom, we first discuss the concepts of upper and lower bounds.

Definition 1.3. Let S be a non-empty subset of $\mathbb{R}$.

(1) The set S is **bounded above** if there exists a real number u such that

$$x \le u \quad \text{for all } x \in S.$$

Such an element u is called an **upper bound** of S.

(2) The set S is **bounded below** if there exists a real number l such that

$$l \le x \quad \text{for all } x \in S.$$

Such an element l is called a **lower bound** of S.

(3) The set S is **bounded** if it is both bounded above and bounded below. A set is **unbounded** if it is not bounded.

So S is *not* bounded above if for every real number a, there exists an element $x \in S$ with $x > a$. Likewise, y is *not* an upper bound of S if and only if there exists an element $x \in S$ such that $x > y$. Similarly, S is *not* bounded below if for every real number a, there exists an element $x \in S$ with $x < a$. Finally, y is *not* a lower bound of S if and only if there exists an element $x \in S$ such that $x < y$.

Example 1.7.

(1) The subset

$$S = \left\{ \frac{1}{n} : n \in \mathbb{Z}_+ \right\}$$

of $\mathbb{R}$ is bounded. It is bounded above with 1 serving as an upper bound. Any real number $x \ge 1$ is also an upper bound of S. Also, S is bounded below with 0 serving as a lower bound. Any real number $y \le 0$ is also a lower bound of S. In particular, *upper and lower bounds are not unique.*

(2) The subset

$$T = \{x : x < 0\}$$

of $\mathbb{R}$ is bounded above with 0 as an upper bound. However, T is not bounded below.

(3) The subset

$$U = \{x : x > 1\}$$

of $\mathbb{R}$ is bounded below with 1 as a lower bound. However, U is not bounded above.

As noted above, upper and lower bounds, when they exist, are not unique. Thus, when a subset S of $\mathbb{R}$ is bounded above, it makes sense to ask if there is an upper bound u that is in some sense the most efficient one. In other words, is there an upper bound u of S such that no real number $x < u$ is an upper bound of S? To make this precise, we need the following definition.

Definition 1.4. Let S be a non-empty subset of $\mathbb{R}$.

(1) Suppose that S is bounded above. A real number u is called a **supremum**, or **least upper bound**, of S if

- u is an upper bound of S, and
- if $x < u$, then x is not an upper bound of S.

We denote a supremum of S by $\sup(S)$.

(2) Suppose that S is bounded below. A real number l is called an **infimum**, or **greatest lower bound**, of S if

- l is a lower bound of S, and
- if $x > l$, then x is not a lower bound of S.

We denote an infimum of S by $\inf(S)$.

Note that at this point, we do not really know if a set S that is bounded above has a supremum or not. However, if a supremum of S exists, then it must be unique (Exercise (10)). The same is true for infimum, so we can speak of *the* supremum and *the* infimum of S.

Example 1.8. Consider the subset $S = \{x : x < 0\}$ of $\mathbb{R}$. It is bounded above with 0 as an upper bound. It is intuitively clear that 0 should be the least upper bound as well. To prove this, pick any real number $y < 0$. We want to show that y is not an upper bound of S. It suffices to demonstrate the existence of an element $x \in S$ with $x > y$. Since $y < 0$, we have $y < \frac{y}{2} < 0$, and so $\frac{y}{2} \in S$ shows that y is not an upper bound of S. This shows that $\sup(S) = 0$. Observe that the supremum 0 is *not* an element in S.

1.3.4 *The Completeness Axiom*

In this book, we will take the following property of $\mathbb{R}$ for granted.

The Completeness Axiom. Every non-empty subset S of $\mathbb{R}$ that is bounded above has a supremum.

So the Completeness Axiom of $\mathbb{R}$ guarantees the existence of the supremum as long as the non-empty set is bounded above. Although this axiom is stated only for sets that are bounded above, there is a corresponding statement for sets that are bounded below. It states that *every non-empty subset S of $\mathbb{R}$ that is bounded below has an infimum* (Exercise (9b) below). The Completeness Axiom is extremely important because many results later depend on it.

We will now discuss some consequences of the Completeness Axiom. Suppose that every step that you take covers a distance of, say, two feet. If you want to get from town A to town B by walking, it seems obvious that you can accomplish this, provided that you walk long enough. This is the motivation for the following result.

Theorem 1.2 (The Archimedean Property). *Let a and b be positive real numbers. Then there exists a positive integer n such that $na > b$.*

Proof. This is proved by contradiction. Suppose that $na \leq b$ for every positive integer n. So the set
$$S = \{na : n = 1, 2, \ldots\}$$
is not empty, since it contains a, and is bounded above with b as an upper bound. By the Completeness Axiom, the set S has a supremum u. Since $a > 0$, it follows that $u - a < u$. With u being the least upper bound of S, this implies that $u - a$ is not an upper bound of S. So there exists an element $na \in S$ with $na > u - a$, or equivalently, $(n+1)a > u$. But $n + 1$ is a positive integer, so
$$(n+1)a \in S \quad \text{and} \quad (n+1)a \leq u$$
because u is an upper bound of S. This contradicts the statement $(n+1)a > u$. Therefore, there exists a positive integer n with $na > b$. $\qquad\square$

Given a real number $a > 0$, it seems clear that the reciprocal of some large positive integer n lies in between 0 and a. The following result shows that this is indeed the case.

Corollary 1.1. *For every positive real number a, there exists a positive integer n such that $\frac{1}{n} < a$.*

Proof. Applying the Archimedean Property to the positive real numbers a and 1, we obtain a positive integer n such that $na > 1$. The desired inequality now follows when we divide by n. $\qquad\square$

Another consequence of the Archimedean Property is that the set $\mathbb{N}$ of natural numbers is not bounded above.

Corollary 1.2. *Let x be a real number. Then there exists a positive integer n such that $x < n$.*

Proof. If $x \leq 0$, then we can take $n = 1$. So suppose that $x > 0$. Applying the Archimedean Property to 1 and x, we obtain a positive integer n such that $n \cdot 1 = n > x$. $\qquad\square$

It seems obvious that every positive real number is either a natural number or lies between two natural numbers. This is proved in the following result.

Corollary 1.3. *Let x be a positive real number. Then there exists a positive integer n such that*
$$n - 1 \leq x < n.$$

Proof. By Corollary 1.2 there exists at least one positive integer $m > x$. So the set

$$S = \{m \in \mathbb{Z}_+ : m > x\}$$

is non-empty. Let n be the least positive integer in S. Then we have $n > x$. Also, $n - 1$ is a natural number that is strictly less than n, so $n - 1 \notin S$ and $n - 1 \leq x$. Thus, we have $n - 1 \leq x < n$. □

1.3.5 *Lots of Rationals and Irrationals*

Next we want to establish the fact that given any two distinct real numbers x and y, there exist a rational number a and an irrational number b, both of which lie strictly between x and y. This is again a consequence of the Archimedean Property. First we prove the rational case.

Theorem 1.3. *Let x and y be real numbers with $x < y$. Then there exists a rational number a such that*

$$x < a < y.$$

Proof. We prove this when $x > 0$. The other case $x \leq 0$ is Exercise (20). Since $y - x > 0$, there exists a positive integer n satisfying $\frac{1}{n} < y - x$ by Corollary 1.1. Rearranging this inequality we obtain

$$1 + nx < ny.$$

Since $nx > 0$, by Corollary 1.3 there exists a positive integer m such that

$$m - 1 \leq nx < m.$$

This implies that

$$m \leq 1 + nx < ny.$$

Thus, we have $nx < m < ny$. Dividing by n, we obtain

$$x < \frac{m}{n} < y,$$

which proves the theorem. □

For the irrational case, we will need the following result, which shows that $\sqrt{2}$ is an irrational number.

Theorem 1.4. *The real number $\sqrt{2}$ is irrational.*

Proof. This is proved by contradiction. Suppose that $\sqrt{2} = \frac{m}{n}$ for some integers m and n with $n \neq 0$. Since we can always cancel the common integer factors of m and n, we may assume that m and n do not have any common factor > 1. Squaring the equality $\sqrt{2}n = m$, we obtain $2n^2 = m^2$. This shows that 2 is a factor of m^2 and hence of m. It follows that m^2 has at least two factors of 2. The above equality then tells us that n^2 must have at least one factor of 2 as well, so n also has a factor of 2. Thus, 2 is a common factor of m and n, which is a contradiction. Therefore, $\sqrt{2}$ is an irrational number. □

Corollary 1.4. *Let x and y be real numbers with $x < y$. Then there exists an irrational number b such that*

$$x < b < y.$$

Proof. Applying Theorem 1.3 to $\sqrt{2}x < \sqrt{2}y$, we obtain a rational number a such that $\sqrt{2}x < a < \sqrt{2}y$. Dividing by $\sqrt{2}$, we obtain

$$x < \frac{a}{\sqrt{2}} < y.$$

Observe that $b = \frac{a}{\sqrt{2}}$ is an irrational number. Indeed, if b is rational, then since a is rational, so is $\sqrt{2} = \frac{a}{b}$, contradicting Theorem 1.4. □

1.3.6 Intervals

Many results that we will discuss in later chapters have to do with intervals, which we define formally below. As you already know, if an interval contains two points, then any point in between is also in the interval.

Definition 1.5. By an **interval** in $\mathbb{R}$ we mean a non-empty subset I containing at least two real numbers such that if s and t are both in I with $s < t$, then any real number x satisfying $s < x < t$ is also an element in I.

For real numbers a and b with $a < b$, it is easy to see that the following sets are intervals:

(1) $(a,b) = \{x : a < x < b\}$ (open bounded interval).
(2) $(a,b] = \{x : a < x \le b\}$ (half-open bounded interval).
(3) $[a,b) = \{x : a \le x < b\}$ (half-open bounded interval).
(4) $[a,b] = \{x : a \le x \le b\}$ (closed bounded interval).
(5) $(a,\infty) = \{x : a < x\}$ (open unbounded interval).
(6) $[a,\infty) = \{x : a \le x\}$ (closed unbounded interval).
(7) $(-\infty,a) = \{x : x < a\}$ (open unbounded interval).
(8) $(-\infty,a] = \{x : x \le a\}$ (closed unbounded interval).
(9) $(-\infty,\infty) = \mathbb{R}$ (both open and closed interval).

In fact, an interval must be of one of the above forms. We consider the examples of bounded intervals.

Proposition 1.1. *Let I be an interval that is bounded. Then there exist some real numbers a and b such that $I = (a,b)$, $(a,b]$, $[a,b)$, or $[a,b]$.*

Proof. Since I is bounded above and below, it has a supremum b and an infimum a by the Completeness Axiom. Note that a and b may or may not be elements in I. To prove the assertion, it suffices to prove that every real number x satisfying $a < x < b$ is an element in I.

Pick any positive real number ϵ with $\epsilon < \min\{x - a, b - x\}$. Note that we have

$$a + \epsilon < x < b - \epsilon.$$

Since $a + \epsilon > a$, there exists an element $s \in I$ such that

$$s < a + \epsilon < x.$$

Likewise, since $b - \epsilon < b$, there exists an element $t \in I$ such that

$$t > b - \epsilon > x.$$

So we have $s < x < t$ with s and t in I. By the definition of an interval, we conclude that x is an element in I, as desired. $\qquad\square$

In Exercise (24) below you are asked to show that an unbounded interval must be of one of the remaining five forms.

1.3.7   *Exercises*

(1) Prove that the set $\mathbb{Q}$ of rational numbers, with the usual addition and multiplication, is a field. Are $\mathbb{N}$ and $\mathbb{Z}$ fields?

(2) This exercise is about the field with two elements. Consider the set $\mathbb{Z}/2 = \{0, 1\}$ with two elements. Define its addition and multiplication by

$$0 + 0 = 0,\ 0 + 1 = 1 = 1 + 0,\ 1 + 1 = 0,\ 0 \cdot 0 = 0,\ 0 \cdot 1 = 0 = 1 \cdot 0,\ 1 \cdot 1 = 1.$$

   (a) Prove that $\mathbb{Z}/2$ is a field.
   (b) Is $\mathbb{Z}/2$ an ordered field?

(3) Let a and b be real numbers.

   (a) Prove that $||a| - |b|| \le |a - b|$.
   (b) Prove that $|a - b| \le |a| + |b|$.

(4) Suppose that $|x - a| < \frac{\epsilon}{2}$ and that $|y - a| < \frac{\epsilon}{2}$. Prove that $|x - y| < \epsilon$.

(5) Prove that $x = 0$ if and only if the inequality $|x| < \epsilon$ holds for every $\epsilon > 0$.

(6) Suppose that $x < y + \epsilon$ for every $\epsilon > 0$. Prove that $x \le y$.

(7) Prove that $|x - a| < \epsilon$ if and only if x lies in the open interval $(a - \epsilon, a + \epsilon)$.

(8) Prove that a non-empty subset S of $\mathbb{R}$ is bounded if and only if there exists a positive real number M such that $|x| < M$ for every element x in S.

(9) Let S be a non-empty subset of $\mathbb{R}$ that is bounded below.

   (a) Define the set $-S = \{-x : x \in S\}$. Prove that $-S$ is bounded above.
   (b) Prove that $\sup(-S) = -\inf(S)$, i.e., $-\sup(-S)$ is the greatest lower bound of S. Conclude that a non-empty subset of $\mathbb{R}$ that is bounded below has an infimum.

(10) Let S be a non-empty subset of $\mathbb{R}$.

   (a) If S is bounded above, prove that S has a unique supremum.
   (b) If S is bounded below, prove that S has a unique infimum.

(11) Let S be a set consisting of n elements, where n is some positive integer.

    (a) Prove that S is bounded.

    (b) Prove that $\inf(S)$ and $\sup(S)$ are both elements in S.

(12) Let S be a non-empty bounded subset of $\mathbb{R}$.

    (a) Let $u \in \mathbb{R}$ be an upper bound of S. Prove that $u = \sup(S)$ if and only if for every real number $\epsilon > 0$ there exists an element $x \in S$ such that $u - \epsilon < x$.

    (b) Let $l \in \mathbb{R}$ be a lower bound of S. Prove that $l = \inf(S)$ if and only if for every real number $\epsilon > 0$ there exists an element $y \in S$ such that $l + \epsilon > y$.

(13) Let S be a non-empty bounded subset of $\mathbb{R}$. Prove that $\inf(S) \le \sup(S)$.

(14) Let $S \subseteq T$ be non-empty bounded subsets of $\mathbb{R}$. Prove the inequalities:

$$\inf(T) \le \inf(S) \le \sup(S) \le \sup(T).$$

So the infimum of a bigger set is smaller, and the supremum of a bigger set is bigger.

(15) Let S and T be non-empty bounded subsets of $\mathbb{R}$ such that $x \le y$ for all x in S and y in T. Prove the following statements.

    (a) $\sup(S) \le \inf(T)$.

    (b) $\sup(S) = \inf(T)$ if and only if, for every $\epsilon > 0$, there exist x in S and y in T such that $y - x < \epsilon$.

(16) Let S and T be non-empty bounded subsets of $\mathbb{R}$.

    (a) Prove that $\inf(S \cup T) = \min\{\inf(S), \inf(T)\}$, the minimum of $\inf(S)$ and $\inf(T)$.

    (b) Prove that $\sup(S \cup T) = \max\{\sup(S), \sup(T)\}$, the maximum of $\sup(S)$ and $\sup(T)$.

(17) Let a and b be real numbers with $a < b$.

    (a) Find the infimum and the supremum of the open interval $I = (a, b)$.

    (b) Do $\inf(I)$ and $\sup(I)$ lie in I?

    (c) Repeat the previous two parts with the closed interval $[a, b]$.

(18) Let S be a non-empty subset of $\mathbb{R}$ that is bounded above with supremum b. Suppose that b is not an element in S. Prove that for every $\epsilon > 0$, there exist infinitely many elements in S that lie in the interval $(b - \epsilon, b)$.

(19) Find the infimum and supremum of each of the following sets of real numbers.

    (a) $S = \{x : 1 < x \le 3\}$.

    (b) $S = \{x : x^2 - 2x - 3 < 0\}$.

    (c) $S = \{x : x^2 - 5 < 0\}$.

(20) Finish the proof of Theorem 1.3 by proving the case $x \le 0$.

(21) Let $a < b$ be real numbers.

    (a) Prove that there exist infinitely many rational numbers in the interval (a, b).

    (b) Prove that there exist infinitely many irrational numbers in the interval (a, b).

(22) Prove that the following real numbers are all irrational: $\sqrt{3}$, $\sqrt[3]{2}$, and $\sqrt{2}+\sqrt{3}$.

(23) Let I and J be two intervals such that the intersection $I \cap J$ contains at least two distinct real numbers. Prove that $I \cap J$ is also an interval.

(24) Let I be an interval that is an unbounded set.

    (a) If I is bounded below, prove that I has the form (a, ∞) or $[a, \infty)$.

    (b) If I is bounded above, prove that I has the form $(-\infty, a)$ or $(-\infty, a]$.

    (c) If I is neither bounded above nor bounded below, prove that $I = \mathbb{R}$.

(25) In each case, write down an explicit bijection between the given sets.

    (a) $(1, 3)$ and $(5, 12)$.

    (b) $\mathbb{R}$ and $(0, 1)$.

    (c) $[0, 1)$ and $[1, \infty)$.

(26) Describe each of the following sets: $\bigcap_{n=1}^{\infty}[1 - \frac{1}{n}, 1]$, $\bigcap_{n=1}^{\infty}(1 - \frac{1}{n}, 1)$, and $\bigcap_{n=1}^{\infty}[n, \infty)$.

1.4   Mathematical Induction

Many proofs in this book and elsewhere use a technique called *mathematical induction*. This method allows one to establish the validity of many related statements at the same time.

1.4.1   *Induction*

We begin by stating a property of the set $\mathbb{Z}_+$ of positive integers. If S is a non-empty subset of $\mathbb{Z}_+$, then a **least element** of S is an element $x \in S$ such that $x \le y$ for all elements $y \in S$.

Well-Ordering Property. Any non-empty subset of $\mathbb{Z}_+$ has a least element.

This seems rather obvious. If S is a non-empty subset of $\mathbb{Z}_+$, just pick its least element! In any case, we take the Well-Ordering Property as an axiom.

Principle of Mathematical Induction. For each positive integer n, let $P(n)$ be a statement. Suppose that the following two conditions hold:

(1) $P(1)$ is true.

(2) For each positive integer k, if $P(k)$ is true, then $P(k + 1)$ is true.

Then $P(n)$ is true for all positive integers n.

Proof. Let S be the subset of $\mathbb{Z}_+$ consisting of those positive integers n for which $P(n)$ is not true. If S is the empty-set, then we are done. Otherwise, S is not empty. By the Well-Ordering Property there is a least element N in S. Since $P(1)$

is true, we have that $1 \notin S$ and that $N > 1$. So $N - 1$ is a positive integer that does not lie in S, which means that $P(N-1)$ is true. By the second hypothesis above, it follows that

$$P(N) = P((N-1)+1)$$

is true. So $N \notin S$, which is a contradiction. Therefore, S must be empty, and $P(n)$ is true for all positive integers n. □

In some situations, the statements $P(n)$ do not start with $n = 1$ but some other positive integer n_0. In this case, the initial case $P(1)$ should be replaced by $P(n_0)$ and k should be replaced by $k \geq n_0$.

We illustrate how to use induction with a few examples.

Theorem 1.5 (Bernoulli's Inequality). *Suppose that $a > -1$ and $a \neq 0$. Then for integers $n > 1$, we have*

$$(1+a)^n > 1 + na.$$

Proof.   This is proved by induction on n. For the first case, when $n = 2$, we have

$$(1+a)^2 = 1 + 2a + a^2 > 1 + 2a.$$

Now suppose that $(1+a)^n > 1 + na$ for some $n \geq 2$. For the case $n + 1$, we have

$$(1+a)^{n+1} = (1+a)^n(1+a) > (1+na)(1+a)$$
$$= 1 + (n+1)a + na^2 > 1 + (n+1)a,$$

as desired. □

For a positive integer n, define n **factorial** as the product

$$n! = 1 \cdot 2 \cdot 3 \cdots (n-1)n.$$

For example, $1! = 1$, $3! = 6$, and $5! = 120$. We set $0! = 1$. For non-negative integers n and k with $n \geq k$, define the **binomial coefficient** as

$$\binom{n}{k} = \frac{n!}{k!(n-k)!}. \tag{1.4}$$

For example,

$$\binom{n}{0} = 1, \quad \binom{n}{n} = 1, \quad \binom{4}{2} = 6, \quad \text{and} \quad \binom{5}{2} = 10.$$

The binomial coefficients are related as follows. The following result is proved without induction. It is needed for the result after it. Its proof is a simple algebraic computation, which is left to the reader as an exercise.

Theorem 1.6 (Pascal's Triangle). *For positive integers n and k with $n \geq k$, we have*

$$\binom{n}{k} + \binom{n}{k-1} = \binom{n+1}{k}.$$

Theorem 1.7. *For a real number x and a positive integer n, we have*

$$(1+x)^n = \sum_{k=0}^{n} \binom{n}{k} x^k.$$

Proof. This is proved by induction on n. The first case is

$$(1+x)^1 = 1 + x = \binom{1}{0} x^0 + \binom{1}{1} x^1.$$

Suppose that the required identity is true for some positive integer n. We must show that the next case is true. We compute as follows:

$$(1+x)^{n+1} = (1+x) \cdot (1+x)^n$$

$$= (1+x) \cdot \sum_{k=0}^{n} \binom{n}{k} x^k$$

$$= \sum_{k=0}^{n} \binom{n}{k} x^k + \sum_{k=0}^{n} \binom{n}{k} x^{k+1}$$

$$= 1 + \sum_{k=1}^{n} \binom{n}{k} x^k + \sum_{k=1}^{n} \binom{n}{k-1} x^k + \binom{n}{n} x^{n+1}$$

$$= 1 + \sum_{k=1}^{n} \left(\binom{n}{k} + \binom{n}{k-1} \right) x^k + x^{n+1}$$

$$= \binom{n+1}{0} x^0 + \sum_{k=1}^{n} \binom{n+1}{k} x^k + \binom{n+1}{n+1} x^{n+1}$$

$$= \sum_{k=0}^{n+1} \binom{n+1}{k} x^k.$$

In the next-to-the-last step, we used Theorem 1.6. The theorem is proved. $\square$

Corollary 1.5 (Binomial Theorem). *For real numbers a and b and a positive integer n, we have*

$$(a+b)^n = \sum_{k=0}^{n} \binom{n}{k} a^k b^{n-k}.$$

Proof. If $b = 0$ then the identity is clearly true, since $n - k > 0$ except for $n = k$. So suppose that $b \neq 0$. In this case, we have

$$(a+b)^n = b^n \left(1 + \frac{a}{b}\right)^n$$

$$= b^n \sum_{k=0}^{n} \binom{n}{k} \left(\frac{a}{b}\right)^k$$

$$= \sum_{k=0}^{n} \binom{n}{k} a^k b^{n-k},$$

where we used Theorem 1.7 in the second equality. $\square$

1.4.2   *Strong Induction*

Sometimes the following form of induction is more convenient to use.

Principle of Strong Induction. For each positive integer n, let $P(n)$ be a statement. Suppose that the following two conditions hold:

(1) $P(1)$ is true.
(2) For each positive integer k, if $P(1), \ldots, P(k)$ are true, then $P(k+1)$ is true.

Then $P(n)$ is true for all positive integers n.

Proof.   This is proved by contradiction. Suppose that $P(n)$ is not true for some n. By the Well-Ordering Property we can choose the least N such that $P(N)$ is not true. Since $P(1)$ is true, we have that $N > 1$ and that $P(1), \ldots, P(N-1)$ are true. By the second hypothesis it follows that $P(N)$ is true, which is a contradiction. Thus, $P(n)$ is true for every positive integer n.   □

1.4.3   *Exercises*

(1) Prove that $\sum_{k=1}^{n} k = \frac{n(n+1)}{2}$ for every positive integer n.

(2) Prove that $\sum_{k=1}^{n} k^2 = \frac{n(n+1)(2n+1)}{6}$ for every positive integer n.

(3) Prove that $\sum_{k=1}^{n} k^3 = \left(\frac{n(n+1)}{2}\right)^2$.

(4) Prove that $\sum_{k=1}^{n} (8k-5) = 4n^2 - n$ for every positive integer n.

(5) Prove that $\sum_{k=1}^{n} (2k-1)^2 = \frac{n(4n^2-1)}{3}$ for every positive integer n.

(6) Prove that $\sum_{k=0}^{n} x^k = \dfrac{1 - x^{n+1}}{1 - x}$ for every real number $x \neq -1$ and positive integer n.

(7) Prove that $\binom{n}{k} = \binom{n}{n-k}$ for all integers $n \geq k \geq 0$.

(8) Prove that $7^n - 2^n$ is divisible by 5 for every positive integer n.

(9) Prove that $11^n - 4^n$ is divisible by 7 for every positive integer n.

(10) Prove that $5^{2n} - 1$ is divisible by 8 for every positive integer n.

(11) Prove that $\sum_{k=1}^{n} \frac{1}{2^k} < 1$ for every positive integer n.

(12) Prove that $2^n < n!$ for all integers $n \geq 4$.

(13) Prove that $n^2 < 2^n$ for all integers $n \geq 5$.

(14) Prove that $n^3 \leq 3^n$ for every positive integer n.

(15) Let $a_1, \ldots, a_n$ be real numbers with $n \geq 2$.

    (a) Prove that $|a_1 + \cdots + a_n| \leq |a_1| + \cdots + |a_n|$.

    (b) Prove that $|a_1 + \cdots + a_n| \geq |a_1| - (|a_2| + \cdots + |a_n|)$.

(16) Let $a_1, \ldots, a_n$ be positive real numbers with $n \geq 1$. Prove the inequality:
$$(1 + a_1) \cdots (1 + a_n) \geq 1 + a_1 + \cdots + a_n.$$

(17) Let $S_1, \ldots, S_n$ be non-empty bounded subsets of $\mathbb{R}$ for some $n \geq 1$. Prove that the union $\bigcup_{i=1}^{n} S_i$ is also non-empty and bounded.

(18) Prove Theorem 1.6.

1.5   Countability

Sets come in different sizes. The most basic way to define the size of a set is to categorize it as either finite or infinite.

1.5.1   *Finite and infinite sets*

Definition 1.6. A set S is called a **finite set** if either S is the empty set $\varnothing$ or there is a bijection from S to the set $\{1, 2, \ldots, n\}$ with n elements for some positive integer n. A set that is not finite is called an **infinite set**.

If there is a bijection $f: S \to T$, then its inverse function $f^{-1}: T \to S$ is also a bijection (Exercise (13) on page 7). Thus, in the above definition, there is a bijection from S to the set $\{1, 2, \ldots, n\}$ if and only if there is a bijection from $\{1, 2, \ldots, n\}$ to S.

Example 1.9. The set $\mathbb{N}$ of natural numbers is infinite, which seems obvious. To prove it, suppose to the contrary that there exists a bijection $f: \{1, 2, \ldots, n\} \to \mathbb{N}$ for some positive integer n. We can rank the n natural numbers $f(1), \ldots, f(n)$ from the smallest to the largest. If $f(k)$ is the largest integer among these n natural numbers, then $f(k) + 1 \in \mathbb{N}$ is not in the range of f. This contradicts the assumption that f is a bijection. Therefore, the set $\mathbb{N}$ is infinite.

It should be intuitively clear that any set that contains an infinite set is itself an infinite set. This is made precise in the following result.

Theorem 1.8. *Let T be an infinite subset of a set S. Then S is an infinite set.*

Proof.   We prove this by contradiction. So suppose that S is not an infinite set, which means that S is a finite set. By definition there is a bijection $f: \{1, \ldots, n\} \to S$ for some positive integer n. We will derive the contradiction that T is a finite set. Let $k_1, \ldots, k_i$ be the list of all the distinct positive integers in $\{1, \ldots, n\}$ such that $f(k_1), \ldots, f(k_i)$ are elements in T. In particular, we have

$$\{f(k_1), \ldots, f(k_i)\} = T.$$

The function $g: \{1, \ldots, i\} \to T$ defined by

$$g(j) = f(k_j) \in T$$

is injective because f is injective. Moreover, g is surjective because every element in T is of the form $f(k_j)$ for some $j = 1, \ldots, i$. Thus, g is a bijection, which means that T is a finite set. This is a contradiction, and hence S is an infinite set.   □

Corollary 1.6. *Let T be a subset of a finite set S. Then T is a finite set.*

Proof.   This is the contrapositive of Theorem 1.8. Indeed, if T is an infinite set, then S is an infinite set as well, contradicting the assumption.   □

Corollary 1.7. *The sets $\mathbb{Z}$, $\mathbb{Q}$, and $\mathbb{R}$ are all infinite.*

Proof. We know from Example 1.9 that $\mathbb{N}$ is an infinite set. Since $\mathbb{N}$ is a subset of $\mathbb{Z}$, $\mathbb{Q}$, and $\mathbb{R}$, it follows from Theorem 1.8 that they are also infinite sets. $\quad\square$

1.5.2   *Countable and Uncountable Sets*

What is perhaps surprising is that there are very different sizes even among infinite sets. To make this precise, we need the following definitions.

Definition 1.7. An infinite set S is said to be **countable** if there is a bijection from $\mathbb{Z}_+$ to S. An infinite set that is not countable is called **uncountable**.

So a set S is uncountable if and only if it is neither finite nor countable. An uncountable set is in some sense a lot bigger than a countable set. At this point, it may seem hard to imagine an infinite set that is uncountable. We will address this issue in a short while. Before that we want to show that the set $\mathbb{Q}$ of rational numbers is countable. This might also seem counter-intuitive because $\mathbb{Q}$ contains many elements that are not in $\mathbb{Z}_+$.

Theorem 1.9. *The set $\mathbb{Q}$ of rational numbers is countable.*

Proof. Every rational number can be written uniquely as a fraction $\frac{m}{n}$ in lowest terms, in which m and n are integers, $n > 0$, and that m and n do not have common integer factors > 1. Therefore, we can create a list of the rational numbers as follows:

$$0 \quad 1 \quad -1 \quad 2 \quad -2 \quad 3 \quad -3 \ldots$$

$$\tfrac{1}{2} \quad -\tfrac{1}{2} \quad \tfrac{3}{2} \quad -\tfrac{3}{2} \quad \tfrac{5}{2} \quad -\tfrac{5}{2} \quad \tfrac{7}{2} \ \cdots$$

$$\tfrac{1}{3} \quad -\tfrac{1}{3} \quad \tfrac{2}{3} \quad -\tfrac{2}{3} \quad \tfrac{4}{3} \quad -\tfrac{4}{3} \quad \tfrac{5}{3} \ \cdots$$

$$\tfrac{1}{4} \quad -\tfrac{1}{4} \quad \tfrac{3}{4} \quad -\tfrac{3}{4} \quad \tfrac{5}{4} \quad -\tfrac{5}{4} \quad \tfrac{7}{4} \ \cdots$$
$$\vdots \qquad\qquad\quad \ddots \qquad\qquad\quad \vdots$$

In this list the nth row contains the rational numbers $\frac{m}{n}$ in lowest terms whose denominator is n. Each rational number appears in this list exactly once.

    Now we define a function $f: \mathbb{Z}_+ \to \mathbb{Q}$ by going down the southwest-to-northeast diagonals. In other words, define

First diagonal $f(1) = 0$,

Second diagonal $f(2) = \dfrac{1}{2}$, $f(3) = 1$,

Third diagonal $f(4) = \dfrac{1}{3}$, $f(5) = -\dfrac{1}{2}$, $f(6) = -1$,

and so forth. This function f is surjection because every rational number appears in the above list. Also, f is injective because every rational number appears only once in the list. Therefore, f is a bijection from $\mathbb{Z}_+$ to $\mathbb{Q}$. This shows that $\mathbb{Q}$ is countable. □

Next we observe that the set $\mathbb{R}$ is uncountable, so intuitively $\mathbb{R}$ is a lot bigger as a set than $\mathbb{Q}$. We will concentrate on the open interval $(0,1)$.

Theorem 1.10. *The set of real numbers in the open interval $(0,1)$ is uncountable.*

Ideas of Proof. The plan is to show that, if $(0,1)$ is countable, then there is a real number in this interval whose nth digit in its decimal expansion differs from that in the nth term in the countable set $(0,1)$. This would then be a contradiction.

Proof. This is proved by contradiction. Suppose that the set $(0,1)$ is countable, so there is a bijection $f\colon\mathbb{Z}_+ \to (0,1)$. Each real number in $(0,1)$ has a decimal expansion

$$0.x_1x_2x_3x_4\cdots$$

in which each digit x_i is between 0 to 9. Using the bijection f, we can list all the elements in $(0,1)$ as follows:

$$f(1) = 0.a_{11}a_{12}a_{13}a_{14}\cdots,$$
$$f(2) = 0.a_{21}a_{22}a_{23}a_{24}\cdots,$$
$$f(3) = 0.a_{31}a_{32}a_{33}a_{34}\cdots,$$
$$\vdots \qquad \ddots$$

Each digit a_{ij} is between 0 and 9.

To derive a contradiction, we will write down explicitly a real number b in $(0,1)$ that is different from $f(n)$ for every $n \in \mathbb{Z}_+$. Consider the digits along the diagonal: a_{11}, a_{22}, a_{33}, a_{44}, etc. For each integer $n \geq 1$ define the digit

$$b_n = \begin{cases} 2 & \text{if } a_{nn} \neq 2, \\ 5 & \text{if } a_{nn} = 2. \end{cases}$$

Consider the real number

$$b = 0.b_1b_2b_3\cdots \in (0,1).$$

For each integer $n \geq 1$, b is different from $f(n) = 0.a_{n1}a_{n2}a_{n3}\cdots$ because their nth digits, b_n and a_{nn}, are different. This shows that f is not surjective, contradicting the assumption that f is a bijection. Therefore, $(0,1)$ is an uncountable set. □

The technique used in the proof above is called **Cantor's diagonal argument**.

1.5.3 *Exercises*

(1) Let $S_1, \ldots, S_k$ be finite sets for some positive integer k.

    (a) Prove that the finite union $\bigcup_{i=1}^{k} S_i$ is also a finite set.

    (b) Prove the the finite Cartesian product $\prod_{i=1}^{k} S_i$ is also a finite set.

(2) Let S_n, where $n \in \mathbb{Z}_+$, be an infinite collection of sets.

    (a) Is the union $\bigcup_{n=1}^{\infty} S_n$ necessarily an infinite set?

    (b) Is the Cartesian product $\prod_{n=1}^{\infty} S_n$ necessarily an infinite set?

(3) Let S be an infinite set, and let A be a finite set. Prove that $S \setminus A$ is an infinite set.

(4) Prove that the set $\mathbb{N}$ is countable.

(5) Let A be a set, and let B be a countable set. Prove that A is countable if and only if there exists a bijection from A to B.

(6) Let S be an infinite set. Prove that S has a countable subset.

(7) Prove that the set $\mathbb{Z}$ of integers is countable by writing down an explicit bijection from $\mathbb{Z}_+$ to $\mathbb{Z}$.

(8) Let $n \geq 2$ be a positive integer. Prove that $\mathbb{Z}$ is the disjoint union of n countable sets.

(9) Write down a bijection from the set of even integers to the set $\mathbb{Z}$ of integers.

(10) Let T be a subset of a countable set S. Prove that T is either finite or countable.

(11) Let S and T be countable sets. Prove that there is a bijection from S to $S \cup T$.

(12) Let $f : T \to S$ be an injection in which S is countable. Prove that T is either finite or countable.

(13) Let S be a countable set, and let T be a set. Suppose that there is a surjection $f : S \to T$. Prove that T is either finite or countable.

(14) Let $\{S_n\}$ be an infinite collection of countable sets, where $n \in \mathbb{Z}_+$. Prove that the union $\bigcup_{n=1}^{\infty} S_n$ is a countable set.

(15) Prove that the Cartesian product $\mathbb{Z}_+ \times \mathbb{Z}_+$ is a countable set by writing down an explicit bijection from $\mathbb{Z}_+$ to $\mathbb{Z}_+ \times \mathbb{Z}_+$.

(16) Let S be a countable set, and let T be a finite subset of S. Prove that $S \setminus T$ is countable.

(17) Let S be an uncountable set, and let T be a countable set. Prove that $S \setminus T$ is uncountable.

(18) Let S be a set that contains an uncountable subset. Prove that S is uncountable.

(19) Let S be a set, and let T be an uncountable set. Prove that $S \cup T$ is uncountable.

(20) Let S be a set, and let T be an uncountable set. Suppose that there is a surjection $f : S \to T$. Prove that S is uncountable.

(21) (a) Prove that the set $\mathbb{R}$ of real numbers is uncountable.

    (b) Prove that the set $\mathbb{I}$ of irrational numbers is uncountable.

(22) Let $f : S \to T$ be a bijection in which S is uncountable. Prove that T is

uncountable.

(23) Let a and b be real numbers with $a < b$.

    (a) Write down an explicit bijection $f : (0, 1) \to (a, b)$.

    (b) Prove that the interval (a, b) is an uncountable set.

(24) Let a and b be real numbers with $a < b$.

    (a) Prove that there exist countably many rational numbers in (a, b).

    (b) Prove that there exist uncountably many irrational numbers in (a, b).

1.6   Additional Exercises

(1) Let $f : S \to T$ be an injection with inverse function $f^{-1} : Ran(f) \to Dom(f) = S$. For any subset A of $Ran(f)$, there seem to be two meanings of the symbol $f^{-1}(A)$: (i) the inverse image of A under f, or (ii) the image of A under f^{-1}. Convince yourself that these two interpretations give rise to the same subset of S.

(2) Let $S_n = \{n - 1, n\}$ for positive integers n. Prove that $\bigcup_{n=1}^{\infty} S_n = \mathbb{N}$.

(3) Consider the subset $\mathbb{Q}(\sqrt{2}) = \{a + b\sqrt{2} : a, b \in \mathbb{Q}\}$ of $\mathbb{R}$. Prove that $\mathbb{Q}(\sqrt{2})$ is a field in which the addition and multiplication are the ones in $\mathbb{R}$.

(4) Let S be a non-empty bounded subset of $\mathbb{R}$, and let a be a real number. Define the sets $a + S = \{a + x : x \in S\}$ and $aS = \{ax : x \in S\}$.

    (a) Prove that $a + S$ and aS are both non-empty and bounded.

    (b) Prove that $\inf(a + S) = a + \inf(S)$ and $\sup(a + S) = a + \sup(S)$.

    (c) If $a > 0$, prove that $\inf(aS) = a\inf(S)$ and $\sup(aS) = a\sup(S)$.

    (d) If $a < 0$, prove that $\inf(aS) = a\sup(S)$ and $\sup(aS) = a\inf(S)$.

(5) Let $\{I_n\}$ be an infinite collection of intervals, where $n = 1, 2, \ldots$. Suppose that the intersection $I = \bigcap_{i=1}^{\infty} I_i$ contains at least two distinct real numbers. Prove that I is an interval.

(6) Let $S_1, \ldots, S_n$ be countable sets for some positive integer n. Prove that the Cartesian product $S_1 \times \cdots \times S_n$ is a countable set.

(7) Let S be an uncountable set. Prove that there is a *proper* subset T of S such that there is a bijection from S to T.

(8) Let S be an uncountable set, and let T be a countable set. Prove that there is a bijection from S to $S \cup T$.

(9) Let S be the set of functions from $\mathbb{Z}_+$ to the set $\{0, 1\}$ with two elements. Prove that S is uncountable.

(10) Let S and T be two sets. Their **symmetric difference** is defined as the set

$$S \bigtriangleup T = \{x : x \in S \text{ or } x \in T \text{ and } x \notin S \cap T\}.$$

    (a) Prove that $S \bigtriangleup T = (S \smallsetminus T) \cup (T \smallsetminus S)$.

    (b) Prove that $S = (S \bigtriangleup T) \cup (S \cap T)$, where the union is disjoint.

    (c) Prove that $S \cap (T \bigtriangleup U) = (S \cap T) \bigtriangleup (S \cap U)$.

(d) Prove that $(S \cup T) \mathbin{\triangle} (U \cup V) \subseteq (S \mathbin{\triangle} U) \cup (T \mathbin{\triangle} V)$. Does equality hold in general?

(11) In this exercise we construct a bijection $f : (0,1] \to (0,1)$. For each positive integer n, consider the function defined as $f_n(x) = \frac{3}{2^n} - x$ whose domain is the interval $(\frac{1}{2^n}, \frac{1}{2^{n-1}}]$. Define $f : (0,1] \to (0,1)$ by setting $f(x) = f_n(x)$ when $x \in (\frac{1}{2^n}, \frac{1}{2^{n-1}}]$.

    (a) Sketch the graphs of f_1, f_2, f_3, and f_4.

    (b) Prove that f_n is a bijection from its domain to the interval $[\frac{1}{2^n}, \frac{1}{2^{n-1}})$.

    (c) Conclude that f is a bijection.

(12) By adapting the ideas of the previous exercise or otherwise, construct a bijection between $(0,1)$ and $[0,1)$.

(13) Prove that there exist pairwise-disjoint intervals I_n, where $n = 1, 2, 3, \ldots$, such that the closed interval $[0,1]$ is the union $\bigcup_{n=1}^{\infty} I_n$. Here **pairwise-disjoint** means $I_i \cap I_j = \varnothing$ whenever $i \neq j$.

(14) A subset S of $\mathbb{R}$ is said to be **open** if for every element $x \in S$, there exists $\epsilon > 0$ such that the open interval $(x - \epsilon, x + \epsilon)$ is a subset of S. Say that S is **closed** if $\mathbb{R} \setminus S$ is open.

    (a) Prove that $\varnothing$ and $\mathbb{R}$ are both open and closed.

    (b) Exhibit a subset S of $\mathbb{R}$ that is neither open nor closed.

(15) Let a and b be real numbers with $a < b$.

    (a) Prove that the intervals (a,b), (a,∞), and $(-\infty,a)$ are open.

    (b) Prove that the intervals $[a,b]$, $[a,\infty)$, and $(-\infty,a]$ are closed.

(16) Let $S_1, S_2, \ldots$ be subsets of $\mathbb{R}$.

    (a) Suppose that each S_n is open. Prove that the union $\bigcup_{i=1}^{\infty} S_i$ is also open.

    (b) Suppose that each S_n is closed. Prove that the intersection $\bigcap_{i=1}^{\infty} S_i$ is also closed.

(17) Let $S_1, S_2, \ldots$ be *open* subsets of $\mathbb{R}$.

    (a) Prove that the *finite* intersection $S_1 \cap \cdots \cap S_n$ is open.

    (b) Is it necessarily true that the intersection $\bigcap_{i=1}^{\infty} S_i$ is open?

(18) Let $S_1, S_2, \ldots$ be *closed* subsets of $\mathbb{R}$.

    (a) Prove that the *finite* union $S_1 \cup \cdots \cup S_n$ is closed.

    (b) Is it necessarily true that the union $\bigcup_{i=1}^{\infty} S_i$ is closed?

(19) Let S be a set. Prove that there is an injection from S to $\mathcal{P}(S)$. Recall that the **power set** $\mathcal{P}(S)$ of a set S is the set whose elements are the subsets of S.

(20) Let S and T be two sets.

    (a) Suppose that there is a bijection $f : S \to T$. Prove that there is a bijection from $\mathcal{P}(S)$ to $\mathcal{P}(T)$.

    (b) If $\mathcal{P}(S) = \mathcal{P}(T)$, is it true that $S = T$?

(21) Let S be a set. Prove that there is *no* surjection from S to $\mathcal{P}(S)$.

(22) Prove that $\mathcal{P}(\mathbb{Z}_+)$ is uncountable.

(23) For a set S, let $\mathcal{P}_k(S)$ be the set of subsets of S with exactly k elements, where k is an arbitrary positive integer.

   (a) Prove that there is a bijection from S to $\mathcal{P}_1(S)$.
   (b) If S has exactly n elements for some positive integer n, prove that $\mathcal{P}_k(S)$, where $k \le n$, has $\binom{n}{k}$ elements.

(24) Let k be a positive integer.

   (a) Prove that $\mathcal{P}_k(\mathbb{Z}_+)$ is countable.
   (b) Prove that the set of all *finite* subsets of $\mathbb{Z}_+$ is countable.

(25) For each positive integer n, let $I_n = [a_n, b_n]$ be a bounded closed interval. Suppose that $I_{n+1} \subseteq I_n$ for each $n \ge 1$.

   (a) Prove that the set $A = \{a_n : n \ge 1\}$ is bounded above.
   (b) Prove that the set $B = \{b_n : n \ge 1\}$ is bounded below.
   (c) Let $\alpha = \sup(A)$ and $\beta = \inf(B)$. Prove that $\alpha \le \beta$ and that both α and β lie in $\bigcap_{n=1}^{\infty} I_n$.
   (d) Prove that $[\alpha, \beta] = \bigcap_{n=1}^{\infty} I_n$. If $\alpha = \beta$, then $[\alpha, \beta]$ means the set $\{\alpha\}$. In particular, the intersection $\bigcap_{n=1}^{\infty} I_n$ is non-empty. This is called the **Nested Interval Property**.

Chapter 2

Sequences

In this chapter we discuss sequences of real numbers, which the reader should have encountered briefly in calculus. Given a sequence, the main question is whether it converges to a limit or not. A related question concerns the convergence of subsequences. The concepts of limits and convergence will occur many more times later in this book.

In section 2.1 the concept of a convergent sequence is introduced. In section 2.2 some basic properties of limits are discussed, including preservation of arithmetic operations and certain types of inequalities. A basic but very important result, the Monotone Convergence Theorem, is proved.

Section 2.3 is about the Bolzano-Weierstrass Theorem, which guarantees the existence of a convergent subsequence of a bounded sequence. Using this result, we give another characterization of a convergent sequence as a Cauchy sequence. With the concept of Cauchy sequences, one can establish the convergence of a sequence without first knowing what its limit is. Similar Cauchy criteria for convergence will occur again later in this book.

In section 2.4 we discuss limit superior and limit inferior. These concepts provide more information about the possible limits of subsequences of a given sequence. They will also be used in the next chapter when we discuss series.

2.1 Sequences of Real Numbers

2.1.1 *Definition of a Sequence*

A sequence is a list of real numbers. We will define a sequence formally below. Before that let us consider a few examples.

Example 2.1.

(1) The sequence

$$\left\{1, \frac{1}{2}, \frac{1}{3}, \frac{1}{4}, \dots\right\}$$

has $a_n = \frac{1}{n}$ as its nth term.

(2) The sequence

$$\{-1, 1, -1, 1, \ldots\}$$

has $a_n = (-1)^n$ as its nth term.

(3) The sequence

$$\{a, a^2, a^3, a^4, \ldots\}$$

has $a_n = a^n$ as its nth term, where a is some fixed real number.

(4) For $a > 0$ the sequence

$$\left\{a, a^{\frac{1}{2}}, a^{\frac{1}{3}}, a^{\frac{1}{4}}, \ldots\right\}$$

has as its nth term $a_n = a^{\frac{1}{n}}$.

(5) The sequence

$$\left\{(1+1)^1, \left(1+\frac{1}{2}\right)^2, \left(1+\frac{1}{3}\right)^3, \ldots\right\}$$

has as its nth term $a_n = (1+\frac{1}{n})^n$.

(6) The *Fibonacci sequence*

$$\{1, 1, 2, 3, 5, 8, 13, 21, 34, \ldots\}$$

is defined recursively as

$$a_1 = a_2 = 1 \quad \text{and} \quad a_n = a_{n-1} + a_{n-2} \quad \text{for} \quad n \geq 3.$$

The nth term in the Fibonacci sequence is called the nth *Fibonacci number*. In other words, for $n \geq 3$ the nth Fibonacci number is the sum of the previous two Fibonacci numbers.

(7) For a real number r, there is a sequence $\{s_n\}$ whose nth term is the sum

$$s_n = 1 + r + r^2 + \cdots + r^n.$$

This sequence is usually called a *geometric series*.

In the first example $\{\frac{1}{n}\}$, if n is large, then its reciprocal $\frac{1}{n}$ is close to 0. The larger n gets, the closer $\frac{1}{n}$ is to 0. Intuitively this tells us that the sequence $\{\frac{1}{n}\}$ converges to 0. The only problem is that we do not yet have a rigorous definition of convergence. This will be given in a short while.

Definition 2.1. A **sequence** is a function from the set $\mathbb{Z}_+$ of positive integers to $\mathbb{R}$.

If $f: \mathbb{Z}_+ \to \mathbb{R}$ is such a function, we will usually write $f(n)$ as a_n and call the list

$$\{a_n\} = \{a_1, a_2, \ldots\}$$

a sequence. Since f and $\{a_n\}$ determine each other, these two ways of representing a sequence are equivalent.

Occasionally, we may have a sequence starting with some a_k instead of a_1. For example, if $a_n = (n-1)^{-1}$ then we cannot take $n = 1$. In such cases, we write

$$\{a_n\}_{n=k}^{\infty} \quad \text{or} \quad \{a_n\}_{n \geq k}$$

for the sequence $\{a_k, a_{k+1}, a_{k+2}, \ldots\}$.

2.1.2 Convergent Sequences

For a sequence $\{a_n\}$ to converge to some real number L, what it should mean is this: The sequence $\{a_n\}$ can get as close to L as one wants, as long as one is allowed to discard a few terms in $\{a_n\}$. Closeness here means distance from L. So if $\epsilon > 0$ is some small positive real number, being close to L means being an element in the interval $(L - \epsilon, L + \epsilon)$. In other words, given any $\epsilon > 0$, all but a finite number of terms a_n should lie in $(L - \epsilon, L + \epsilon)$.

In formal mathematical language, this is phrased as follows.

Definition 2.2. Let $\{a_n\}$ be a sequence, and let L be a real number. We say that the sequence $\{a_n\}$ **converges to** L if for every $\epsilon > 0$, there exists a positive integer N such that

$$n \geq N \quad \text{implies} \quad |a_n - L| < \epsilon. \tag{2.1}$$

In this case, we say that L is the **limit** of the sequence and that $\{a_n\}$ is **convergent**. If $\{a_n\}$ converges to L, we will also write

$$\lim a_n = L, \quad \lim_{n \to \infty} a_n = L, \quad \text{or} \quad a_n \to L,$$

and vice versa. A sequence that is not convergent is called **divergent**.

Observe that in Definition 2.2, *first* $\epsilon > 0$ is given, and *then* N is chosen to make (2.1) true. The condition (2.1) means that

$$|a_N - L| < \epsilon, \quad |a_{N+1} - L| < \epsilon, \quad |a_{N+2} - L| < \epsilon,$$

and so forth. In other words, the a_n are within ϵ of L for all n sufficiently large.

Here is the negation of convergence: A sequence $\{a_n\}$ *does not converge* to L if and only if there exists $\epsilon_0 > 0$ such that, for every positive integer N, there exists an integer

$$n \geq N \quad \text{such that} \quad |a_n - L| \geq \epsilon_0.$$

Observe that this n is dependent on N.

Example 2.2. As discussed above, it should be the case that $\frac{1}{n} \to 0$. To prove this, let $\epsilon > 0$ be given. We want to make the inequality

$$\left| \frac{1}{n} - 0 \right| < \epsilon$$

true for all n sufficiently large. By Corollary 1.2 there exists a positive integer $N > \frac{1}{\epsilon}$. Then for any integer $n \geq N$, we have

$$\left| \frac{1}{n} - 0 \right| = \frac{1}{n} \leq \frac{1}{N} < \epsilon.$$

This shows that $\{\frac{1}{n}\}$ converges to 0.

Example 2.3. The sequence $\{(-1)^n\} = \{-1, 1, -1, 1, \ldots\}$ should not converge to any real number L, since it keeps alternating between 1 and -1. To prove that it is divergent, let L be any real number, and we show that $\{(-1)^n\}$ does not converge to L. Consider $\epsilon_0 = \frac{1}{2}$. For any positive integer N, either

$$|1 - L| \geq \frac{1}{2} \quad \text{or} \quad |-1 - L| \geq \frac{1}{2}.$$

In fact, if neither one of these inequalities is true, then by the Triangle Inequality (Theorem 1.1),

$$2 = |1 - (-1)| = |(1 - L) + (L - (-1))| \leq |1 - L| + |L - (-1)| < \frac{1}{2} + \frac{1}{2} = 1,$$

which is absurd. Therefore, the sequence $\{(-1)^n\}$ is divergent.

Example 2.4. For $0 < a < 1$ consider the sequence with $a_n = a^{\frac{1}{n}}$. With the example $a = \frac{1}{2}$, one can guess that $a^{\frac{1}{n}} \to 1$. To prove this, consider $b_n > 0$ defined by the equality

$$a^{\frac{1}{n}} = \frac{1}{1 + b_n},$$

which is possible because $0 < a^{\frac{1}{n}} < 1$. We need an estimate of $|1 - a^{\frac{1}{n}}|$. Observe that

$$\left|1 - a^{\frac{1}{n}}\right| = \left|1 - \frac{1}{1 + b_n}\right| = \left|\frac{b_n}{1 + b_n}\right| < b_n,$$

since $1 + b_n > 1$. Thus, it suffices to get an estimate for b_n. Rearranging its definition, we have

$$1 = 1^n = a(1 + b_n)^n > a(1 + nb_n)$$

by Bernoulli's Inequality (Theorem 1.5). Since both a and b_n are positive, we can rearrange the above inequality to obtain

$$b_n < \frac{1 - a}{an}.$$

Suppose that $\epsilon > 0$ is given. We want to show that $|1 - a^{\frac{1}{n}}| < \epsilon$ for all n sufficiently large. By Corollary 1.2 there exists a positive integer N such that

$$N > \frac{1 - a}{a\epsilon}.$$

Then for $n \geq N$, we have

$$\left|1 - a^{\frac{1}{n}}\right| < b_n < \frac{1 - a}{an} \leq \frac{1 - a}{aN} < \epsilon.$$

This shows that $a^{\frac{1}{n}} \to 1$ when $0 < a < 1$.

Example 2.5. For $|a| < 1$ consider the sequence $\{a^n\}$. If $a = 0$, then $a^n = 0$ for every n, and we have $a^n \to 0$. If $a \neq 0$, observe that

$$|a| > |a|^2 > |a|^3 > \cdots.$$

With the example $a = \frac{1}{2}$, it is not hard to guess that the sequence $\{a^n\}$ should converge to 0. To prove this, let $\epsilon > 0$ be given. We can write

$$|a| = \frac{1}{1+b}$$

for some $b > 0$. We estimate $|a^n - 0| = |a|^n$ as follows:

$$|a^n - 0| = \frac{1}{(1+b)^n} < \frac{1}{1+nb} < \frac{1}{nb},$$

where the first inequality uses the Bernoulli's Inequality (Theorem 1.5). By Corollary 1.2 there exists a positive integer $N > \frac{1}{b\epsilon}$. Then for $n \geq N$ we have

$$|a^n - 0| < \frac{1}{nb} \leq \frac{1}{Nb} < \epsilon.$$

This shows that $a^n \to 0$ for $|a| < 1$.

2.1.3 *Divergent Sequences*

In Example 2.5, suppose that we consider the sequence $\{a^n\}$ with $a > 1$ instead. In this case, we have $0 < \frac{1}{a} < 1$, so $\frac{1}{a^n} \to 0$. This means that a^n is large when n is large. In fact, a^n can be as large as one wants, provided that n is sufficiently large. We make this precise in the following definition.

Definition 2.3. Let $\{a_n\}$ be a sequence.

(1) We say that $\{a_n\}$ **diverges to** ∞ if for every real number M, there exists a positive integer N such that

$$n \geq N \quad \text{implies} \quad a_n > M.$$

In this case, we write

$$\lim a_n = \infty \quad \text{or} \quad a_n \to \infty,$$

and vice versa.

(2) We say that $\{a_n\}$ **diverges to** $-\infty$ if for every real number M, there exists a positive integer N such that

$$n \geq N \quad \text{implies} \quad a_n < M.$$

In this case, we write

$$\lim a_n = -\infty \quad \text{or} \quad a_n \to -\infty,$$

and vice versa.

A sequence $\{a_n\}$ *does not diverge to* ∞ if and only if there exists a real number M_0 such that for every positive integer N, there exists an integer

$$n \geq N \quad \text{such that} \quad a_n \leq M_0.$$

We leave it to the reader to formulate what it means for a sequence to *not* diverge to $-\infty$. The reader should be careful that a divergent sequence does not necessarily diverge to ∞ or $-\infty$, as the following example illustrates.

Example 2.6. We saw in Example 2.3 that the sequence $\{(-1)^n\}$ is divergent. However, it does not diverge to ∞ or $-\infty$. In fact, pick $M_0 = 1$. Then

$$a_n = (-1)^n \leq M_0$$

for all n, so $\{(-1)^n\}$ does not diverge to ∞. A similar argument shows that $\{(-1)^n\}$ does not diverge to $-\infty$.

Example 2.7. Recall the Fibonacci sequence $\{1, 1, 2, 3, 5, 8, 13, 21, 34, \ldots\}$ with

$$a_1 = a_2 = 1 \quad \text{and} \quad a_n = a_{n-1} + a_{n-2}$$

for $n \geq 3$. It appears that $\{a_n\}$ diverges to ∞. To prove this, let M be a real number. We want to show that $a_n > M$ for all n sufficiently large. Looking at the Fibonacci numbers, it should be clear that $a_n \geq n$ for $n \geq 5$. We prove this by induction, with the first case being $a_5 = 5$. The second case is $a_6 = 8 > 6$. Suppose that $a_k \geq k$ for $5 \leq k \leq n$ for some $n \geq 6$. Then we have

$$a_{n+1} = a_n + a_{n-1} \geq n + (n-1) > n + 1.$$

Therefore, by induction we conclude that $a_n \geq n$ for all $n \geq 5$. Thus, we can choose N to be $\max\{5, M+1\}$. Then for $n \geq N$ we have

$$a_n \geq n \geq N > M,$$

showing that the Fibonacci sequence diverges to ∞.

2.1.4   *Uniqueness of Limits*

If a sequence $\{a_n\}$ is convergent, is it possible that there are two distinct limits? If M and L are two limits of this sequence, then the a_n are all close to L for all n sufficiently large. But the a_n are also close to M for all n sufficiently large. This suggests that the two limits L and M must, in fact, be equal. This is proved precisely in the next result.

Theorem 2.1. *Let $\{a_n\}$ be a convergent sequence. Then it has a unique limit.*

Proof.   Suppose that

$$\lim a_n = L \quad \text{and} \quad \lim a_n = M$$

for some real numbers L and M. We want to show that $L = M$. It is enough to show that $|L - M|$ can be made arbitrarily small. By the Triangle Inequality (Theorem 1.1), we have

$$|L - M| = |(L - a_n) + (a_n - M)| \le |L - a_n| + |a_n - M|. \tag{2.2}$$

So for an arbitrary $\epsilon > 0$, to ensure that $|L - M| < \epsilon$, it is enough to have the right-hand side of (2.2) to be less than ϵ. Since $\lim a_n = L$, there exists a positive integer N_1 such that

$$n \ge N_1 \quad \text{implies} \quad |a_n - L| < \frac{\epsilon}{2}.$$

Likewise, since $\lim a_n = M$, there exists a positive integer N_2 such that

$$n \ge N_2 \quad \text{implies} \quad |a_n - M| < \frac{\epsilon}{2}.$$

Thus, taking $N = \max\{N_1, N_2\}$, we have that $n \ge N$ implies

$$|L - M| \le |L - a_n| + |a_n - M| < \frac{\epsilon}{2} + \frac{\epsilon}{2} = \epsilon. \tag{2.3}$$

Since $\epsilon > 0$ is arbitrary, this shows that $L = M$. $\qquad\square$

The argument used in the proof of Theorem 2.1 is called an $\frac{\epsilon}{2}$-**argument**. This type of argument and its variants will appear many more times in this book. In this type of argument, the quantity to be estimated, $|L - M|$ in the case above, is split into two parts as in (2.2). An estimate of $\frac{\epsilon}{2}$ is obtained for each of the two parts. Then the results are combined as in (2.3) to get the desired estimate. Variations of this argument (e.g., an $\frac{\epsilon}{3}$-argument) involve splitting the quantity to be estimated into three or more parts.

2.1.5 *Exercises*

(1) In Definition 2.2 suppose that we replace "$n \ge N$" by "$n > N$" in (2.1). Show that the resulting definition of convergence is equivalent to the original one.

(2) Prove the following statements using the definition of convergence.

(a) $2 + \dfrac{3}{n} \to 2$.

(b) $\dfrac{5 - \sqrt{n}}{n} \to 0$.

(c) $\sqrt{3 - \dfrac{7}{n}} \to \sqrt{3}$.

(d) $4 + \dfrac{13}{\sqrt[3]{n}} \to 4$.

(e) $\dfrac{2^n}{n!} \to 0$, where $n! = 1 \cdot 2 \cdots (n-1) \cdot n$.

(3) Prove the following statements using the definition of convergence.

(a) $\dfrac{1 - 4n}{n^2} \to 0$.

(b) $\dfrac{3 - 2n}{1 + 5n} \to \dfrac{-2}{5}$.

(c) $\dfrac{2n^2 + 7}{3n^2 - n - 1} \to \dfrac{2}{3}$.

(d) $\dfrac{n^2 - 15}{4n^2 + 17} \to \dfrac{1}{4}$.

(4) Prove the following statements using Definition 2.3.

    (a) $n^2 - 2n - 10 \to \infty$.

    (b) $3^n \to \infty$.

    (c) $\dfrac{n!}{3^n} \to \infty$.

    (d) $-\ln(n) \to -\infty$.

(5) Let a be a real number. Prove that the sequence with $a_n = a$ for all n is convergent with limit a.

(6) Prove that $(\sqrt{n+1} - \sqrt{n}) \to 0$.

(7) Let $a > 1$ be a real number. Prove that $a^{\frac{1}{n}} \to 1$.

(8) Prove that $a_n \to L$ if and only if $(a_n - L) \to 0$.

(9) Prove that $a_n \to L$ if and only if $-a_n \to -L$.

(10) Prove that $a_n \to L$ if and only if for every $\epsilon > 0$, the interval $(L-\epsilon, L+\epsilon)$ contains all but a finite number of the a_n.

(11) Prove that $a_n \to 0$ if and only if $|a_n| \to 0$.

(12) If $a_n \to L$, prove that $|a_n| \to |L|$. Is the converse true?

(13) Suppose that $a_n \to L$ for some real number $L > 0$. Prove that there exists a positive integer N such that $n \geq N$ implies $a_n > 0$.

(14) Suppose that $a_n \geq b$ for each n and that $a_n \to L$.

    (a) Prove that $L \geq b$.

    (b) If the hypothesis $a_n \geq b$ is replaced by $a_n > b$, is it necessarily true that $L > b$?

(15) Suppose that $|a_{n+1} - L| < |a_n - L|$ for each integer $n \geq 1$. Does it follow that $a_n \to L$?

(16) Consider the sequences $\{a_n\}_{n=1}^{\infty}$ and $\{a_n\}_{n=k}^{\infty}$, where k is some fixed positive integer > 1. Prove that $\{a_n\}_{n=1}^{\infty}$ converges to L if and only if $\{a_n\}_{n=k}^{\infty}$ converges to L.

(17) Suppose that $a_n \to L$ and $b_n \to L$ (the same L).

    (a) Prove that $(a_n - b_n) \to 0$.

    (b) Prove that $(a_n + b_n) \to 2L$.

(18) Suppose that $|a_n| \leq M$ for all n, where M is some positive real number. Let p be a positive integer. Prove that $\dfrac{a_n}{n^p} \to 0$.

(19) Suppose that $a_n \to 0$ with each $a_n \geq 0$ and that $\{b_n\}$ is a sequence that satisfies $|b_n - L| \leq a_n$ for all $n \geq N$ for some positive integer N.

    (a) Prove that $b_n \to L$.

(b) Suppose that the hypothesis $|b_n - L| \leq a_n$ is replaced by $|b_n - L| \leq ca_n$ for some fixed positive real number c. Prove that $b_n \to L$.

(20) Suppose that $\{a_n\}$ is a convergent sequence, and let $\epsilon > 0$ be a real number.

   (a) Prove that there exists a positive integer N such that $|a_n - a_{n+1}| < \epsilon$ for all $n \geq N$.
   (b) Prove that there exists a positive integer N such that $n, m \geq N$ implies $|a_n - a_m| < \epsilon$.

(21) Let $\{a_n\}$ be a sequence.

   (a) Write down what it means for $\{a_n\}$ to *not* diverge to $-\infty$.
   (b) Give an example of a divergent sequence that does not diverge to $-\infty$.

(22) If $a_n \to \infty$ or $-\infty$, prove that $\{a_n\}$ is divergent.

(23) Suppose that each $a_n > 0$. Prove that $a_n \to 0$ if and only if $\frac{1}{a_n} \to \infty$.

(24) Let $a > 1$ be a real number. Prove that $a^n \to \infty$.

(25) Suppose that $a_n \to \infty$ and $b_n \to \infty$.

   (a) If $c > 0$ is a real number, prove that $ca_n \to \infty$.
   (b) Prove that $(a_n + b_n) \to \infty$.
   (c) Prove that $a_n b_n \to \infty$.

(26) In each case, construct sequences $\{a_n\}$ and $\{b_n\}$ with the given properties.

   (a) $a_n \to \infty$, $b_n \to \infty$, and $(a_n - b_n) \to \infty$.
   (b) $a_n \to \infty$, $b_n \to \infty$, and $(a_n - b_n) \to 17$.
   (c) $a_n \to \infty$, $b_n \to \infty$, $b_n \neq 0$ for each n, and $\frac{a_n}{b_n} \to \infty$.
   (d) $a_n \to \infty$, $b_n \to \infty$, $b_n \neq 0$ for each n, and $\frac{a_n}{b_n} \to 17$.

(27) Suppose that $a_n \to L$ and that $b_n \to \infty$. Prove that $(a_n + b_n) \to \infty$.

2.2   Properties of Limits

In this section we discuss several basic but important properties of sequences, including boundedness and the Monotone Convergence Theorem 2.3.

2.2.1   *Bounded Sequences*

If a sequence $\{a_n\}$ converges to a limit L, then all but a finite number of the a_n are within, say, 1 of L. The other terms, say, $a_1, \ldots, a_{N-1}$, all lie inside a finite interval. Thus, by enlarging this interval to include $(L - 1, L + 1)$, it seems that the entire sequence lies inside some finite closed interval. To make this precise, first we need some definitions.

Definition 2.4. A sequence $\{a_n\}$ is said to be **bounded** if it is bounded as a set (Definition 1.3).

The discussion before the above definition suggests that a convergent sequence is bounded. This is, indeed, the case, as we now show.

Theorem 2.2. *A convergent sequence is bounded.*

Proof.   Let $\{a_n\}$ be a convergent sequence with limit L. Given $\epsilon = 1$, there exists a positive integer N such that

$$n \geq N \quad \text{implies} \quad |a_n - L| < 1.$$

This is equivalent to $a_n \in (L - 1, L + 1)$. Then each a_m satisfies the inequalities

$$\min\{a_1, \ldots, a_{N-1}, L - 1\} \leq a_m \leq \max\{a_1, \ldots, a_{N-1}, L + 1\},$$

showing that the sequence $\{a_n\}$ is bounded. $\square$

The converse of the above theorem is *not* true. In other words, a bounded sequence is not necessarily convergent.

Example 2.8. The sequence $\{(-1)^n\}$ is bounded, since the set $\{-1, 1\}$ is bounded above by 1 and bounded below by -1. However, this sequence is divergent (Example 2.3).

2.2.2   *Monotone Sequences*

The above theorem and example suggest a natural question: If boundedness itself is not enough to guarantee convergence, is there some additional hypothesis on a bounded sequence that can guarantee its convergence? The answer is yes, and the relevant concepts are given in the definitions below.

Definition 2.5. Let $\{a_n\}$ be a sequence.

(1) The sequence $\{a_n\}$ is said to be **increasing** if

$$a_n \leq a_{n+1} \quad \text{for all} \quad n.$$

It is **strictly increasing** if

$$a_n < a_{n+1} \quad \text{for all} \quad n.$$

(2) The sequence $\{a_n\}$ is said to be **decreasing** if

$$a_n \geq a_{n+1} \quad \text{for all} \quad n.$$

It is **strictly decreasing** if

$$a_n > a_{n+1} \quad \text{for all} \quad n.$$

(3) The sequence $\{a_n\}$ is said to be **monotone** if it is either increasing or decreasing. It is **strictly monotone** if it is either strictly increasing or strictly decreasing.

A sequence $\{a_n\}$ is *not* increasing if and only if there exists an integer

$$n_0 \quad \text{such that} \quad a_{n_0} > a_{n_0+1}.$$

It is *not* strictly increasing if there exists an integer

$$n_0 \quad \text{such that} \quad a_{n_0} \geq a_{n_0+1}.$$

It is an exercise for the reader to formulate the negations of *decreasing* and *strictly decreasing*.

Example 2.9.

(1) The sequence $\{(-1)^n\}$ is neither increasing nor decreasing.
(2) The sequence $\{\frac{1}{n}\}$ is strictly decreasing.
(3) The sequence $\{1 - \frac{1}{n}\}$ is strictly increasing.
(4) The sequence $\{1, 1, \frac{1}{2}, \frac{1}{2}, \frac{1}{3}, \frac{1}{3}, \ldots\}$ is decreasing but not strictly decreasing.
(5) The Fibonacci sequence $\{1, 1, 2, 3, 5, 8, 13, \ldots\}$ is increasing but not strictly increasing.

The following result is an answer to the question above.

Theorem 2.3 (Monotone Convergence Theorem). *Let $\{a_n\}$ be a monotone sequence. Then $\{a_n\}$ is convergent if and only if it is bounded. In particular, a monotone bounded sequence is convergent.*

Ideas of Proof. The more precise claim and what the proof below shows is that a bounded increasing sequence $\{a_n\}$ converges to $\sup\{a_n : n \in \mathbb{Z}_+\}$. Likewise, a bounded decreasing sequence converges to $\inf\{a_n : n \in \mathbb{Z}_+\}$.

Proof. If $\{a_n\}$ is convergent, then it is bounded by Theorem 2.2. This proves the "only if" direction.

For the other direction, suppose that $\{a_n\}$ is bounded. We consider the case when $\{a_n\}$ is increasing, leaving the decreasing case to the reader as an exercise. Since the non-empty set

$$A = \{a_1, a_2, \ldots\}$$

is bounded above, it has a supremum α by the Completeness Axiom (section 1.3.4). We show that the sequence converges to $\alpha = \sup(A)$. So let $\epsilon > 0$ be given. Since

$$\alpha - \epsilon < \alpha,$$

$\alpha - \epsilon$ is not an upper bound of the set A. In other words, there exists an a_N such that

$$\alpha - \epsilon < a_N.$$

Then for $n \geq N$, we have

$$\alpha - \epsilon < a_N \leq a_n \leq \alpha < \alpha + \epsilon,$$

from which we obtain

$$|a_n - \alpha| < \epsilon.$$

This proves that $\lim a_n = \alpha$. $\qquad\qquad\qquad\qquad\qquad\qquad\square$

The reader should be careful that the Monotone Convergence Theorem *does not* assert that a convergent sequence is monotone.

Example 2.10. The sequence $\{(-1)^n \frac{1}{n}\} = \{-1, \frac{1}{2}, -\frac{1}{3}, \frac{1}{4}, \ldots\}$ converges to 0. However, it is neither increasing nor decreasing.

Example 2.11. Consider the sequence with $a_n = \sum_{k=0}^{n} \frac{1}{k!}$. We show that it is convergent using the Monotone Convergence Theorem. First, this sequence is strictly increasing, and hence monotone, because

$$a_{n+1} = a_n + \frac{1}{(n+1)!} > a_n.$$

The sequence is bounded below because each a_n is positive. It remains to show that it is bounded above. For $n \geq 2$ we have

$$a_n = 1 + \frac{1}{1!} + \frac{1}{2!} + \frac{1}{3!} + \cdots + \frac{1}{n!}$$
$$\leq 1 + 1 + \frac{1}{2} + \frac{1}{2^2} + \cdots + \frac{1}{2^{n-1}}$$
$$< 2 + 1.$$

This shows that $0 < a_n < 3$ for each n, so the sequence is bounded and monotone. Thus, $\{a_n\}$ is convergent by the Monotone Convergence Theorem.

Example 2.12. Consider the sequence with $a_n = \sum_{k=1}^{n} \frac{1}{k3^k}$. As in the previous example, the a_n are all positive, hence bounded below, and the sequence is strictly increasing. To show that it is bounded above, observe that

$$a_n = \sum_{k=1}^{n} \frac{1}{k3^k} \leq \sum_{k=1}^{n} \frac{1}{3^k} < \sum_{k=1}^{n} \frac{1}{2^k} < 1.$$

Thus, $\{a_n\}$ is a bounded increasing sequence. By the Monotone Convergence Theorem, it is convergent.

2.2.3 *Arithmetics of Sequences*

Next we observe that one can apply the usual arithmetic operations to convergent sequences.

Theorem 2.4. *Suppose* $\lim a_n = a$ *and* $\lim b_n = b$, *and suppose* c *is a real number. Then:*

(1) $\lim(a_n + b_n) = a + b.$
(2) $\lim c a_n = ca.$
(3) $\lim a_n b_n = ab.$
(4) *If* $b_n \neq 0$ *for each* n *and* $b \neq 0$, *then* $\lim \frac{a_n}{b_n} = \frac{a}{b}.$

Proof. We will prove the last part, which is the most difficult one, and leave the first three parts as exercises. So assume that $b_n \neq 0$ for each n and $b \neq 0$. First we show that there exists a positive real number K such that

$$|b_n| > K \quad \text{for all} \quad n.$$

Given $\frac{|b|}{2} > 0$, since $\lim b_n = b$, there exists a positive integer N such that

$$n \geq N \quad \text{implies} \quad |b_n - b| < \frac{|b|}{2}.$$

This implies that

$$|b| - |b_n| < \frac{|b|}{2}$$

by Exercise (3a) on page 15. This in turn is equivalent to

$$\frac{|b|}{2} < |b_n| \quad \text{for} \quad n \geq N.$$

Now

$$K = \frac{1}{2} \min \left\{ |b_1|, \ldots, |b_{N-1}|, \frac{|b|}{2} \right\}$$

satisfies $|b_n| > K$ for all n.

To show that $\lim \frac{a_n}{b_n} = \frac{a}{b}$, we need to estimate

$$\left| \frac{a_n}{b_n} - \frac{a}{b} \right| = \frac{|a_n b - a b_n|}{|b_n b|} = \frac{|a_n(b - b_n) + b_n(a_n - a)|}{|b_n||b|}$$

$$\leq \frac{|a_n|}{|b_n||b|} |b - b_n| + \frac{1}{|b|} |a_n - a| \tag{2.4}$$

where the last step follows from the Triangle Inequality (Theorem 1.1). Given $\epsilon > 0$, we will use an $\frac{\epsilon}{2}$-argument. Since $\lim a_n = a$, there exists a positive integer N_1 such that

$$n \geq N_1 \quad \text{implies} \quad |a_n - a| < \frac{\epsilon |b|}{2}.$$

Moreover, since $\{a_n\}$ is convergent, it is bounded. So there exists a real number $M > 0$ such that $|a_n| < M$ for all n. This implies that

$$\frac{|a_n|}{|b_n||b|} < \frac{M}{K|b|}$$

for all n.

Now, since $\lim b_n = b$, there exists a positive integer N_2 such that

$$n \geq N_2 \quad \text{implies} \quad |b - b_n| < \frac{\epsilon K|b|}{2M}.$$

Thus, for $n \geq N = \max\{N_1, N_2\}$, it follows from (2.4) that

$$\left| \frac{a_n}{b_n} - \frac{a}{b} \right| < \frac{|a_n|}{|b_n||b|} \cdot \frac{\epsilon K|b|}{2M} + \frac{1}{|b|} \cdot \frac{\epsilon |b|}{2} < \frac{M}{K|b|} \cdot \frac{\epsilon K|b|}{2M} + \frac{\epsilon}{2} = \epsilon.$$

This shows that $\lim \frac{a_n}{b_n} = \frac{a}{b}$. $\qquad \square$

The above theorem is very useful in computing limits, as illustrated in the following examples.

Example 2.13.

(1) For any positive integer p, the sequence with $a_n = \frac{1}{n^p}$ converges to 0 because $\frac{1}{n} \to 0$ (Example 2.2) and $a_n = (\frac{1}{n})^p$.

(2) Consider the sequence with

$$a_n = \frac{2n^2 + 7}{3n^2 - n - 1}.$$

Since $\frac{1}{n} \to 0$ and $\frac{1}{n^2} \to 0$, we have

$$a_n = \frac{2n^2 + 7}{3n^2 - n - 1} = \frac{2 + \frac{7}{n^2}}{3 - \frac{1}{n} - \frac{1}{n^2}} \to \frac{2 + 0}{3 - 0 - 0} = \frac{2}{3}.$$

(3) Similarly, the sequence with

$$a_n = \frac{1 - 4n + 3n^2}{n^2}$$

satisfies

$$a_n = \frac{1 - 4n + 3n^2}{n^2} = \frac{1}{n^2} - \frac{4}{n} + 3 \to 0 - 0 + 3 = 3.$$

(4) Let a be a real number with $0 < a < 1$. Then the sequence with

$$a_n = -2a^{\frac{1}{n}} + 5a^n$$

converges to -2 because $a^{\frac{1}{n}} \to 1$ (Example 2.4) and $a^n \to 0$ (Example 2.5).

The next result says that, if a sequence $\{c_n\}$ is bounded term-wise between two sequences, both of which converge to the same limit, then so does $\{c_n\}$.

Theorem 2.5 (Squeeze Theorem). *Suppose that*

$$a_n \le c_n \le b_n \quad for \ all \quad n \quad and \quad \lim a_n = L = \lim b_n$$

for some real number L. Then $\lim c_n = L$.

Proof. This is an $\frac{\epsilon}{3}$-argument. We need to estimate

$$|c_n - L| = |(c_n - a_n) + (a_n - L)| \le |c_n - a_n| + |a_n - L|,$$

where the last inequality uses the Triangle Inequality (Theorem 1.1). So we need to estimate the last two terms above. From the given inequalities we have

$$0 \le c_n - a_n \le b_n - a_n.$$

Let $\epsilon > 0$ be given. Since $\lim a_n = L$ there exists a positive integer N_1 such that

$$n \ge N_1 \quad \text{implies} \quad |a_n - L| < \frac{\epsilon}{3}.$$

Likewise, there exists a positive integer N_2 such that

$$n \geq N_2 \quad \text{implies} \quad |b_n - L| < \frac{\epsilon}{3}.$$

Thus, for $n \geq N = \max\{N_1, N_2\}$, we have

$$|c_n - a_n| \leq |b_n - a_n| \leq |b_n - L| + |L - a_n| < \frac{\epsilon}{3} + \frac{\epsilon}{3} = \frac{2\epsilon}{3}.$$

So for $n \geq N = \max\{N_1, N_2\}$, we have

$$|c_n - L| \leq |c_n - a_n| + |a_n - L| < \frac{2\epsilon}{3} + \frac{\epsilon}{3} = \epsilon.$$

This shows that $\lim c_n = L$. □

Example 2.14. The sequence with $a_n = \frac{\cos n}{n}$ satisfies

$$-\frac{1}{n} \leq a_n \leq \frac{1}{n}$$

because $-1 \leq \cos x \leq 1$ for every real number x. Since both $\frac{1}{n} \to 0$ and $-\frac{1}{n} \to 0$, we conclude that $a_n \to 0$ as well.

2.2.4 *Exercises*

(1) Let $\{a_n\}$ be a sequence.

    (a) Write down what it means for $\{a_n\}$ to *not* be decreasing.
    (b) Write down what it means for $\{a_n\}$ to *not* be strictly decreasing.

(2) Finish the proof of Theorem 2.4 by proving the first three parts.
(3) Prove Theorem 2.3 when $\{a_n\}$ is a bounded decreasing sequence.
(4) Prove that the following sequences converge to 0.

    (a) $a_n = \frac{n!}{n^n}$.
    (b) $a_n = (-1)^n \frac{3}{\sqrt[3]{n+17}}$.
    (c) $a_n = \sqrt{n+17} - \sqrt{n-5}$.

(5) Let $\{a_n\}$ be a monotone sequence.

    (a) If $\{a_n\}$ is not bounded above, prove that $a_n \to \infty$.
    (b) If $\{a_n\}$ is not bounded below, prove that $a_n \to -\infty$.

(6) Let $\{a_n\}$ be a bounded sequence. If $b_n \to 0$, prove that $a_n b_n \to 0$.
(7) Suppose that $a_n \to L$ and $b_n \to M$.

    (a) If $a_n \leq b_n$ for each n, prove that $L \leq M$.
    (b) Given an example in which $a_n < b_n$ for each n and $L = M$.

(8) Suppose that $a \leq c_n \leq b$ for each n and that $c_n \to L$. Prove that $a \leq L \leq b$.
(9) Let $\{a_n\}$ be a sequence with $a_n \geq 0$ for each n. Suppose that $a_n \to L$.

    (a) Prove that $\sqrt{a_n} \to \sqrt{L}$.
    (b) Prove that $\sqrt[3]{a_n} \to \sqrt[3]{L}$.

(10) Let $a_1 = 1$ and $a_{n+1} = \sqrt{1 + a_n}$ for $n \geq 1$.

    (a) Prove that $a_n < 2$ for all n.

    (b) Prove that the sequence $\{a_n\}$ is increasing. Conclude that $\{a_n\}$ is a convergent sequence.

(11) Let $\{a_n\}$ and $\{b_n\}$ be convergent sequences.

    (a) Prove that $\{|a_n + b_n|\}$ is convergent and that

$$\lim |a_n + b_n| \leq \lim |a_n| + \lim |b_n|.$$

    (b) Give an example in which the inequality in the previous part is strict.

    (c) Prove that $\{|a_n b_n|\}$ is convergent and that

$$\lim |a_n b_n| = (\lim |a_n|)\,(\lim |b_n|).$$

(12) Let $a > 0$ be a real number.

    (a) If $0 < r < 1$, prove that $ar^n \to 0$.

    (b) If $r > 1$, prove that $ar^n \to \infty$.

(13) Consider the sequence with $a_n = n^{\frac{1}{n}}$.

    (a) Let $b_n = n^{\frac{1}{n}} - 1$. Using Theorem 1.7 prove that

$$n = (1 + b_n)^n > \frac{1}{2}n(n-1)b_n^2.$$

    (b) Use the previous part to show that $b_n \to 0$.

    (c) Conclude that $n^{\frac{1}{n}} \to 1$.

(14) Prove that $(n + 2)^{\frac{1}{n}} \to 1$.

(15) Let a be a real number with $0 < |a| < 1$.

    (a) Define r by $|a| = \frac{1}{1+r}$. Prove that

$$n|a|^n < \frac{2}{(n-1)r^2}$$

    for every integer $n \geq 2$.

    (b) Prove that $n|a|^n \to 0$.

(16) Let $\{I_n\}$ be a sequence of closed and bounded intervals such that $I_n \supseteq I_{n+1}$ for each n. Use the Monotone Convergence Theorem 2.3 to prove that $\bigcap_{n=1}^{\infty} I_n$ is non-empty. This is called the **Nested Interval Property**.

2.3   The Bolzano-Weierstrass Theorem

The first purpose of this section is to discuss subsequences. The main result is the Bolzano-Weierstrass Theorem 2.8, which asserts that every bounded sequence has a convergent subsequence. This theorem is then used to prove a very useful result due to Cauchy, which gives a necessary and sufficient criterion for convergence.

2.3.1 *Subsequences*

Consider the sequence $\{a_n\} = \{1, \frac{1}{1}, 2, \frac{1}{2}, 3, \frac{1}{3}, \ldots\}$. It neither converges nor diverges to $\pm\infty$. However, if you only pick the first entry and every other entry after that, then the resulting sequence $\{1, 2, 3, \ldots\}$ diverges to ∞. On the other hand, if you only consider the other entries, then the resulting sequence $\{1, \frac{1}{2}, \frac{1}{3}, \ldots\}$ converges to 0. This suggests that something interesting can happen if you only pick certain entries in a sequence.

Definition 2.6. Let $\{a_n\}$ be a sequence. A **subsequence** of $\{a_n\}$ is a sequence $\{a_{n_k}\}$ in which

$$1 \le n_1 \quad \text{and} \quad n_k < n_{k+1}$$

for all $k \ge 1$.

In other words, a subsequence $\{a_{n_k}\}$ is obtained from $\{a_n\}$ by taking only the terms a_{n_1}, a_{n_2}, etc., in this order. Note that the indices n_k have to be strictly increasing, that is, $n_1 < n_2 < n_3 < \cdots$.

Example 2.15.

(1) Given any sequence $\{a_n\}$, $\{a_n\}$ is itself a subsequence, as are $\{a_1, a_3, a_5, \ldots\}$ and $\{a_3, a_6, a_9, \ldots\}$.
(2) The sequence $\{(-1)^n\}$ has $\{1, 1, 1, \ldots\}$ and $\{-1, -1, -1, \ldots\}$ as two subsequences.
(3) For the sequence $\{1, \frac{1}{1}, 2, \frac{1}{2}, 3, \frac{1}{3}, \ldots\}$, $\{2, 1, \frac{1}{2}, \frac{1}{3}, 3, \ldots\}$ is *not* a subsequence.

If a sequence $\{a_n\}$ converges to a limit L, then eventually all the a_n are close to L. So if $\{a_{n_k}\}$ is a subsequence, then the a_{n_k} are all eventually close to L as well. This is made precise in the following result.

Theorem 2.6. *Suppose $\{a_n\}$ converges to L. Then every subsequence $\{a_{n_k}\}$ also converges to L.*

Proof. Given $\epsilon > 0$ there exists a positive integer N such that

$$n \ge N \quad \text{implies} \quad |a_n - L| < \epsilon.$$

Since the indexes n_k are strictly increasing, we can choose an index $n_K \ge N$. Then for $k \ge K$, we have

$$n_k \ge n_K \ge N \quad \text{and} \quad |a_{n_k} - L| < \epsilon.$$

This proves that $\lim a_{n_k} = L$. $\square$

Corollary 2.1. *If $\{a_n\}$ has two convergent subsequences with distinct limits, then $\{a_n\}$ is divergent.*

Proof. Indeed, if $\{a_n\}$ is convergent, then by Theorem 2.6 *every* subsequence converges to the same limit, contradicting the hypothesis. $\square$

Example 2.16. The sequence

$$\{a_n\} = \left\{\frac{1}{1}, 1 - \frac{1}{1}, \frac{1}{2}, 1 - \frac{1}{2}, \frac{1}{3}, 1 - \frac{1}{3}, \ldots\right\}$$

has a subsequence $\{1, \frac{1}{2}, \frac{1}{3}, \ldots\}$, which converges to 0. But it also has another subsequence $\{1 - 1, 1 - \frac{1}{2}, 1 - \frac{1}{3}, \ldots\}$, which converges to 1. Thus the sequence $\{a_n\}$ is divergent by Corollary 2.1.

2.3.2 *Monotone Subsequences*

Most sequences are not monotone. It may come as a surprise that *every* sequence has a monotone subsequence. To prove this result, we need the following concept.

Definition 2.7. Give a sequence $\{a_n\}$, a **peak** is an entry a_n such that

$$a_n \geq a_k \quad \text{for all} \quad k > n.$$

In other words, a peak is an entry that is also an upper bound for all the entries after it.

Theorem 2.7. *Let $\{a_n\}$ be a sequence. Then it has a monotone subsequence.*

Proof. There are two cases. First suppose that the sequence $\{a_n\}$ has infinitely many peaks. If

$$a_{n_1}, a_{n_2}, \ldots$$

are the peaks with $n_1 < n_2 < \cdots$, then the subsequence $\{a_{n_k}\}$ consisting of the peaks is decreasing, and hence monotone.

On the other hand, suppose that $\{a_n\}$ has only finitely many peaks, say,

$$a_{m_1}, \ldots, a_{m_k}.$$

Pick an integer $n_1 > m_k$. Since a_{n_1} is not a peak, there exists an integer

$$n_2 > n_1 \quad \text{such that} \quad a_{n_1} < a_{n_2}.$$

Since $n_2 > n_1 > m_k$, a_{n_2} is not a peak, so there exists an integer

$$n_3 > n_2 \quad \text{such that} \quad a_{n_2} < a_{n_3}.$$

Continuing this way, we obtain an increasing subsequence $\{a_{n_k}\}$. □

We obtain the Bolzano-Weierstrass Theorem by combining the above theorem with the Monotone Convergence Theorem.

Theorem 2.8 (The Bolzano-Weierstrass Theorem). *Every bounded sequence has a convergent subsequence.*

Proof. Let $\{a_n\}$ be a bounded sequence. So there exists a positive real number

$$M \quad \text{such that} \quad |a_n| < M \quad \text{for all} \quad n.$$

By Theorem 2.7 it has a monotone subsequence $\{a_{n_k}\}$. Since

$$|a_{n_k}| < M$$

for all n_k, this subsequence is also bounded. This monotone bounded sequence $\{a_{n_k}\}$ is convergent by the Monotone Convergence Theorem 2.3. ☐

The reader should be careful that an unbounded sequence can have a convergent subsequence as well.

Example 2.17. The sequence $\{a_n\} = \{1, \frac{1}{1}, 2, \frac{1}{2}, 3, \frac{1}{3}, \ldots\}$ is not bounded, but it has a convergent subsequence, namely, $\{\frac{1}{1}, \frac{1}{2}, \frac{1}{3}, \ldots\}$.

2.3.3 *Cauchy Sequences*

Recall that the Monotone Convergence Theorem 2.3 tells us that a monotone bounded sequence $\{a_n\}$ is convergent. In particular, when using this Theorem to establish convergence, we do not need to know in advance what the limit is. We just need to establish that the sequence is bounded and monotone. But what if you have a sequence that is not monotone? Is there a way to tell whether a sequence is convergent without knowing in advance what its limit might be? There is, indeed, such a criterion, which we now discuss.

Definition 2.8. A sequence $\{a_n\}$ is called a **Cauchy sequence** if for every $\epsilon > 0$, there exists a positive integer N such that

$$n, m \geq N \quad \text{implies} \quad |a_n - a_m| < \epsilon.$$

A sequence $\{a_n\}$ is *not* a Cauchy sequence if and only if there exists $\epsilon_0 > 0$ such that for every positive integer N, there exist integers

$$n, m \geq N \quad \text{such that} \quad |a_n - a_m| \geq \epsilon_0.$$

Notice the similarity between this definition and Definition 2.2 for a convergent sequence. However, the above definition does not involve a limit. Roughly speaking, a Cauchy sequence is a sequence whose terms are eventually all close to each other. We are going to show that a sequence is convergent if and only if it is a Cauchy sequence. One direction is in the following result.

Theorem 2.9. *If $\{a_n\}$ is a convergent sequence, then it is a Cauchy sequence.*

Proof. This is an $\frac{\epsilon}{2}$-argument. Let L be the limit of the sequence. Given $\epsilon > 0$ there exists a positive integer N such that

$$n \geq N \quad \text{implies} \quad |a_n - L| < \frac{\epsilon}{2}.$$

Thus, for integers $n, m \geq N$, we have

$$|a_n - a_m| = |(a_n - L) + (L - a_m)| \leq |a_n - L| + |L - a_m| < \frac{\epsilon}{2} + \frac{\epsilon}{2} = \epsilon.$$

Therefore, $\{a_n\}$ is a Cauchy sequence. ☐

To show that Cauchy sequences are convergent, we need to work a little harder. The following observation will be needed.

Lemma 2.1. *Every Cauchy sequence is bounded.*

Proof.   Let $\{a_n\}$ be a Cauchy sequence. Given $\epsilon = 1$ there exists a positive integer N such that

$$n, m \geq N \quad \text{implies} \quad |a_n - a_m| < 1.$$

In particular, for integers $m \geq N$, we have

$$|a_m| - |a_N| \leq |a_N - a_m| < 1.$$

So

$$|a_m| < |a_N| + 1 \quad \text{for all} \quad m \geq N,$$

which implies

$$|a_n| \leq M = \max\{|a_1|, \ldots, |a_{N-1}|, |a_N| + 1\}$$

for all n. Thus, $\{a_n\}$ is a bounded sequence.   □

Using this lemma we can now show that Cauchy sequences are convergent.

Theorem 2.10 (Cauchy Convergence Criterion). *Let $\{a_n\}$ be a Cauchy sequence. Then it is a convergent sequence. Therefore, a sequence is convergent if and only if it is a Cauchy sequence.*

Proof.   This is again an $\frac{\epsilon}{2}$-argument. By Lemma 2.1 $\{a_n\}$ is bounded, so the Bolzano-Weierstrass Theorem 2.8 implies that it has a convergent subsequence $\{a_{n_k}\}$. Let L be the limit of this subsequence. We will show that $\lim a_n = L$. Given $\epsilon > 0$, since $\lim a_{n_k} = L$, there exists a positive integer n_K such that

$$n_k \geq n_K \quad \text{implies} \quad |a_{n_k} - L| < \frac{\epsilon}{2}.$$

Since $\{a_n\}$ is a Cauchy sequence, there exists a positive integer $N \geq n_K$ such that

$$n, m \geq N \quad \text{implies} \quad |a_n - a_m| < \frac{\epsilon}{2}.$$

For integers $m \geq N$, we have

$$n_m \geq N \geq n_K.$$

This implies that

$$|a_m - L| = |(a_m - a_{n_m}) + (a_{n_m} - L)| \leq |a_m - a_{n_m}| + |a_{n_m} - L| < \frac{\epsilon}{2} + \frac{\epsilon}{2} = \epsilon.$$

This shows that the sequence $\{a_n\}$ converges to L.   □

We illustrate how the Cauchy Convergence Criterion can be used in the following examples.

Example 2.18. Consider the sequence with

$$a_n = 1 - \frac{1}{2} + \frac{1}{3} - \cdots + (-1)^{n-1}\frac{1}{n}.$$

We want to show that this sequence is convergent by showing that it is a Cauchy sequence, so we need to estimate $|a_n - a_m|$. If $m > n$, then

$$|a_m - a_n| = \left|(-1)^n \frac{1}{n+1} + (-1)^{n+1}\frac{1}{n+2} + \cdots + (-1)^{m-1}\frac{1}{m}\right|$$

$$= \left|\frac{1}{n+1} - \underbrace{\left(\frac{1}{n+2} - \frac{1}{n+3}\right)}_{\text{positive}} - \underbrace{\left(\frac{1}{n+4} - \frac{1}{n+5}\right)}_{\text{positive}} - \cdots\right|$$

$$\leq \frac{1}{n+1}.$$

Given $\epsilon > 0$ we can choose a positive integer $N > \frac{1}{\epsilon} - 1$. Then for integers $n, m \geq N$ we have

$$|a_m - a_n| \leq \max\left\{\frac{1}{n+1}, \frac{1}{m+1}\right\} \leq \frac{1}{N+1} < \epsilon.$$

This shows that $\{a_n\}$ is a Cauchy sequence. By Theorem 2.10 the sequence $\{a_n\}$ is convergent.

Example 2.19. Consider the sequence with

$$a_n = 1 + \frac{1}{2^2} + \frac{1}{3^2} + \cdots + \frac{1}{n^2}.$$

As in the previous example, we want to show that this is a Cauchy sequence. For $m > n$ we have

$$a_m - a_n = \frac{1}{(n+1)^2} + \frac{1}{(n+2)^2} + \cdots + \frac{1}{m^2}$$

$$< \frac{1}{n(n+1)} + \frac{1}{(n+1)(n+2)} + \cdots + \frac{1}{(m-1)m}$$

$$= \left(\frac{1}{n} - \frac{1}{n+1}\right) + \left(\frac{1}{n+1} - \frac{1}{n+2}\right) + \cdots \left(\frac{1}{m-1} - \frac{1}{m}\right)$$

$$= \frac{1}{n} - \frac{1}{m} < \frac{1}{n}.$$

Given $\epsilon > 0$ choose a positive integer $N > \frac{1}{\epsilon}$. Then for $n, m \geq N$ we have

$$|a_n - a_m| < \max\left\{\frac{1}{n}, \frac{1}{m}\right\} \leq \frac{1}{N} < \epsilon.$$

Thus, $\{a_n\}$ is a Cauchy sequence, hence convergent by Theorem 2.10.

In Example 2.19, in the third line of the calculation we introduced a sum in which pairs of consecutive terms cancel out, leaving only the first and the last terms. This is sometimes referred to as a **telescoping sum**, and the method of proof is called a **telescoping argument**.

2.3.4   *Exercises*

(1) Suppose that $a_n \to \infty$. Prove that every subsequence of $\{a_n\}$ also diverges to ∞. Then prove the statement with $-\infty$ instead of ∞.

(2) Let $\{a_n\}$ and $\{b_n\}$ be two Cauchy sequences. Prove directly from the definition that $\{a_n + b_n\}$ and $\{a_n b_n\}$ are Cauchy sequences.

(3) Prove directly from the definition that the following sequences are Cauchy sequences.

   (a) $a_n = \frac{1}{2+3n}$.
   (b) $a_n = \frac{2n}{5n-7}$.
   (c) $a_n = \frac{3n^2-1}{2n^2+5}$.

(4) Give an example of a divergent sequence $\{a_n\}$ such that $\lim(a_{n+1} - a_n) = 0$.

(5) Let $\{a_n\}$ be a Cauchy sequence such that each a_n is an integer. Prove that there exists a positive integer N such that $\{a_m\}_{m \geq N}$ is a constant sequence.

(6) Let $\{a_n\}$ be a divergent sequence, and let L be a real number. Prove that there exist a positive real number ϵ_0 and a subsequence $\{a_{n_k}\}$ such that $|a_{n_k} - L| \geq \epsilon_0$ for all n_k.

(7) Let $\{a_n\}$ be a bounded sequence, and let L be a real number. Suppose that every convergent subsequence of $\{a_n\}$ converges to L. Prove that $a_n \to L$.

(8) Let $\{a_n\}$ be a Cauchy sequence. Suppose that for every $\epsilon > 0$, there exists an integer $n > \frac{1}{\epsilon}$ such that $|a_n| < \epsilon$. Prove that $a_n \to 0$.

(9) Consider the sequence with

$$a_n = 1 + \frac{1}{2!} + \frac{1}{3!} + \cdots + \frac{1}{n!}.$$

Prove directly from the definition that $\{a_n\}$ is a Cauchy sequence.

(10) Consider the sequence with

$$a_n = 1 + \frac{1}{2} + \frac{1}{3} + \cdots + \frac{1}{n}.$$

Prove that $\{a_n\}$ is not a Cauchy sequence, so $\{a_n\}$ is divergent.

(11) Give an example of a divergent sequence $\{a_n\}$ such that $\lim(a_{n+N} - a_n) = 0$ for every positive integer N.

(12) Let $a_1 = 1$, $a_2 = 2$, and $a_n = \frac{1}{2}(a_{n-1} + a_{n-2})$ for $n \geq 3$.

   (a) For $n \geq 2$ prove that $a_{n+1} - a_n = \left(-\frac{1}{2}\right)^{n-1}$.
   (b) Prove that $\{a_n\}$ is a Cauchy sequence, hence a convergent sequence.

(13) Let a_1 and a_2 be two distinct real numbers. Define $a_n = \frac{1}{2}(a_{n-1} + a_{n-2})$ for $n \geq 3$. Prove that $\{a_n\}$ is a Cauchy sequence.

(14) Let $\{a_n\}$ be a sequence. Suppose that there exists a real number r with $0 < r < 1$ such that $|a_{n+1} - a_n| \leq r^n$ for all n. Prove that $\{a_n\}$ is a Cauchy sequence.

(15) Let $\{a_n\}$ be a sequence and C be a real number such that $0 < C < 1$. Suppose that $|a_{n+1} - a_n| \leq C|a_n - a_{n-1}|$ for all $n \geq 2$. Such a sequence is called a **contractive sequence**.

(a) Prove that $\{a_n\}$ is a Cauchy sequence, hence a convergent sequence.

(b) Is $\{a_n\}$ necessarily a Cauchy sequence if $C = 1$?

(16) Let $\{a_n\}$ be a bounded sequence with $\alpha = \inf\{a_n\}$ and $\beta = \sup\{a_n\}$.

(a) If $\alpha \neq a_n$ for any n, prove that there exists a decreasing subsequence of $\{a_n\}$ that converges to α.

(b) If $\beta \neq a_n$ for any n, prove that there exists an increasing subsequence of $\{a_n\}$ that converges to β.

(17) In each case, construct a sequence $\{a_n\}$ with the given properties.

(a) There exist subsequences converging to 1, 2, and 3.

(b) For every positive integer M, there exists a subsequence converging to M.

(c) There exist subsequences converging to 0 and diverging to ∞ and $-\infty$.

(d) There exist subsequences diverging to ∞ and $-\infty$, and for every positive integer M, there exists a subsequence converging to M.

(18) Let $\{a_n\}$ be a bounded sequence. For each n define $b_n = \sup\{a_k : k \geq n\}$.

(a) Prove that $\{b_n\}$ is a bounded decreasing sequence. In particular, $\{b_n\}$ converges to $L = \inf\{b_n\}$ by the Monotone Convergence Theorem 2.3.

(b) Prove that there exists a subsequence of $\{a_n\}$ that converges to L.

2.4 Limit Superior and Limit Inferior

Theorem 2.6 tells us that, for a convergent sequence, every subsequence has to converge to the same limit. On the other hand, in Example 2.16 we saw that it is possible for a divergent sequence to have convergent subsequences with different limits. A natural question arises. Given a sequence, what can be said about the limits of its convergent subsequences? The purpose of this section is to study this question.

2.4.1 *Subsequential Limits*

First we need some definitions.

Definition 2.9. By an **extended real number** we mean either a real number or one of the symbols ∞ and $-\infty$. The set of extended real numbers is denoted by $\mathbb{R} \cup \{\pm\infty\}$.

The usual order of the real numbers naturally extends to the extended real numbers by defining

$$-\infty < a < \infty \quad \text{for all} \quad a \in \mathbb{R}.$$

Example 2.20. For a monotone sequence $\{a_n\}$, the symbol $\lim a_n$ always makes sense as an extended real number. In fact, if $\{a_n\}$ is bounded, then it is a monotone bounded sequence, which is convergent by the Monotone Convergence Theorem 2.3. If, on the other hand, $\{a_n\}$ is not bounded, then either

$$\lim a_n = \infty \quad \text{or} \quad \lim a_n = -\infty$$

by Exercise (5) on page 43. This discussion still makes sense if the a_n themselves are extended real numbers. For example, if $\{a_n\}$ is increasing and $a_m = \infty$ for some m, then $a_k = \infty$ for all $k \geq m$ and $\lim a_n = \infty$.

Definition 2.10. Let $\{a_n\}$ be a sequence. By a **subsequential limit** of $\{a_n\}$ we mean an extended real number L such that there exists a subsequence $\{a_{n_k}\}$ with $\lim a_{n_k} = L$.

Example 2.21.

(1) If $\{a_n\} \to L$, where $L \in \mathbb{R} \cup \{\pm\infty\}$, then the same is true for every subsequence (Theorem 2.6 and Exercise (1) on page 50). So in this case L is the only subsequential limit of $\{a_n\}$.
(2) The sequence with $a_n = (-1)^n$ has a subsequence $\{1, 1, 1, \ldots\}$ with limit 1 and another subsequence $\{-1, -1, -1, \ldots\}$ with limit -1. Using Exercise (5) on page 50, one can see that ± 1 are the only subsequential limits of $\{a_n\}$.
(3) Consider the sequence

$$\{a_n\} = \{1, 1, 2, 1, 2, 3, 1, 2, 3, 4, \ldots\}.$$

For each positive integer m, there exists a subsequence whose entries are all equal to m. So m is a subsequential limit of $\{a_n\}$. Moreover, there is a subsequence $\{1, 2, 3, 4, 5, \ldots\}$ diverging to ∞. By Exercise (5) on page 50 again, these are the only subsequential limits. So the set of subsequential limits of $\{a_n\}$ is exactly $\mathbb{Z}_+ \cup \{\infty\}$.

Theorem 2.11. *Every sequence has at least one subsequential limit.*

Proof. Given any sequence $\{a_n\}$, it has a monotone subsequence $\{a_{n_k}\}$ by Theorem 2.7. Then $\lim a_{n_k}$ exists as an extended real number (Example 2.20). □

2.4.2   *Size of Subsequential Limits*

Theorem 2.11 tells us that it always makes sense to talk about subsequential limits, whether the sequence itself is convergent or not. We now want to discuss how large or how small the subsequential limits of a sequence can get. First we need some definitions.

Definition 2.11. Let $\{a_n\}$ be a sequence. For each n define the extended real numbers

$$s_n = \sup\{a_k : k \geq n\} \quad \text{and} \quad i_n = \inf\{a_k : k \geq n\}.$$

The **limit superior** of $\{a_n\}$ is defined as the extended real number

$$\limsup a_n = \lim s_n.$$

The **limit inferior** of $\{a_n\}$ is defined as the extended real number

$$\liminf a_n = \lim i_n.$$

If the set $\{a_k : k \geq n\}$ is not bounded above, then we take $s_n = \infty$. Likewise, if the set $\{a_k : k \geq n\}$ is not bounded below, then we take $i_n = -\infty$. The sequence $\{s_n\}$ of extended real numbers is decreasing because the supremum of a smaller set is smaller or stays the same. Similarly, the sequence $\{i_n\}$ is increasing. Therefore, by Example 2.20, $\limsup a_n$ and $\liminf a_n$ always make sense as extended real numbers.

By a sequence we still mean a sequence of real numbers. We will state it explicitly if we consider monotone sequences of extended real numbers.

Example 2.22.

(1) For the sequence with $a_n = (-1)^n$, we have $s_n = 1$ and $i_n = -1$ for every n. Thus, we have

$$\limsup a_n = \lim s_n = 1 \quad \text{and} \quad \liminf a_n = \lim i_n = -1.$$

(2) For the sequence

$$\{a_n\} = \{1, 1, 2, 1, 2, 3, 1, 2, 3, 4, \ldots\}$$

considered in Example 2.21 above, we have $s_n = \sup\{a_k : k \geq n\} = \infty$ and $i_n = \inf\{a_k : k \geq n\} = 1$. Thus, we have

$$\limsup a_n = \infty \quad \text{and} \quad \liminf a_n = 1.$$

Notice that, in this case, $\liminf a_n \leq L \leq \limsup a_n$ for every subsequential limit L of $\{a_n\}$.

In each case of the example above, observe the following:

(1) There exist a subsequence of $\{a_n\}$ converging to $\limsup a_n$ and a subsequence of $\{a_n\}$ converging to $\liminf a_n$. In other words, both $\limsup a_n$ and $\liminf a_n$ are subsequential limits of $\{a_n\}$.
(2) Every subsequential limit L of $\{a_n\}$ satisfies $\liminf a_n \leq L \leq \limsup a_n$. So $\limsup a_n$ and $\liminf a_n$ are, respectively, the largest and the smallest possible subsequential limits of $\{a_n\}$.

We now show that these statements about limit superior and limit inferior are, in fact, true for all sequences of real numbers.

Theorem 2.12. *Let $\{a_n\}$ be a sequence, and let L be a subsequential limit of $\{a_n\}$. Then*

$$\liminf a_n \leq L \leq \limsup a_n.$$

Proof.   Suppose that $\{a_{n_k}\}$ is a subsequence of $\{a_n\}$ such that $\lim a_{n_k} = L$. We first want to show that

$$L \leq \limsup a_n = s.$$

Since $\lim s_n = s$, the same is true for any subsequence of $\{s_n\}$. In particular, we have $\lim s_{n_k} = s$. Since

$$a_{n_k} \leq s_{n_k} \quad \text{for all} \quad k,$$

the inequality is preserved after taking the limits, i.e., $L \leq s$. A similar argument proves the other inequality,

$$\liminf a_n \leq L.$$

We leave the details of this case to the reader as an exercise.   □

2.4.3   *Limit Superior and Limit Inferior are Subsequential Limits*

Theorem 2.13. *Let $\{a_n\}$ be a sequence. Then there exist subsequences $\{a_{n_k}\}$ and $\{a_{n_l}\}$ such that*

$$\lim a_{n_k} = \limsup a_n \quad \text{and} \quad \lim a_{n_l} = \liminf a_n.$$

Ideas of Proof. When $\limsup a_n$ is a real number, the plan is to choose a subsequence of $\{a_n\}$ such that a_{n_k} is close to a suitable s_n. The Squeeze Theorem will then be applied.

Proof.   We will prove the case about limit superior and leave the other case as an exercise for the reader. First consider the case when

$$\limsup a_n = \lim s_n = \infty.$$

We want to construct a subsequence $\{a_{n_k}\}$ of $\{a_n\}$ that diverges to ∞. Since the sequence of extended real numbers $\{s_n\}$ is decreasing, it follows that $s_n = \infty$ for all n. In particular, the set $\{a_k : k \geq 1\}$ is not bounded above, so it is possible to choose an $a_{n_1} > 1$. Next choose an $a_{n_2} > 2$ with $n_2 > n_1$, which is possible because the set $\{a_k : k > n_1\}$ is also not bounded above. Continuing this way, we obtain a subsequence with $a_{n_k} > k$ for each positive integer k. Thus, we have

$$\lim a_{n_k} = \infty = \limsup a_n.$$

Next consider the case when

$$\limsup a_n = -\infty.$$

We know that $a_n \leq s_n$ for each n. Given any real number M, there exists a positive integer N such that

$$n \geq N \quad \text{implies} \quad s_n < M,$$

which in turn implies

$$a_n \le s_n < M.$$

So $\{a_n\}$, as a subsequence of itself, diverges to $-\infty = \lim \sup a_n$.

The only remaining case is when

$$\lim \sup a_n = \lim s_n = L$$

for some real number L. Since $\{s_n\}$ is decreasing, this implies that there exists a positive integer N such that

$$k \ge N \quad \text{implies} \quad s_k < \infty.$$

Since $s_N - 1$ is not an upper bound of $\{a_k : k \ge N\}$, there exists an a_{n_1} with $n_1 \ge N$ such that

$$s_N - 1 < a_{n_1} \le s_N.$$

Next, since $s_{n_1+1} - \frac{1}{2}$ is not an upper bound of $\{a_k : k \ge n_1 + 1\}$, there exists an integer $n_2 > n_1$ such that

$$s_{n_1+1} - \frac{1}{2} < a_{n_2} \le s_{n_1+1}.$$

Continuing this way, we obtain a subsequence $\{a_{n_k}\}$ such that

$$s_{n_{k-1}+1} - \frac{1}{k} < a_{n_k} \le s_{n_{k-1}+1} \tag{2.5}$$

for all $k \ge 2$. Since $s_n \to L$, so does the subsequence $\{s_{n_k+1}\}$ by Theorem 2.6. Taking the limit $\lim_{k \to \infty}$ in (2.5) and using the Squeeze Theorem 2.5, we conclude that $a_{n_k} \to L$, as desired. $\qquad\square$

2.4.4 *Convergence in Terms of Limit Superior and Limit Inferior*

Next we observe that a sequence is convergent, or divergent to $\pm\infty$, if and only if its limit superior and limit inferior coincide.

Theorem 2.14. *Let $\{a_n\}$ be a sequence. Then*

$$\lim a_n = L \in \mathbb{R} \cup \{\pm\infty\} \quad \text{if and only if} \quad \lim \inf a_n = L = \lim \sup a_n.$$

Proof. First consider the "only if" part. It was observed in Example 2.21 (1) that, when $a_n \to L$, the same is true for any subsequence, and L is the only subsequential limit of $\{a_n\}$. By Theorem 2.13 both $\lim \inf a_n$ and $\lim \sup a_n$ are subsequential limits of $\{a_n\}$, so they must both be equal to L.

Next assume that

$$\lim \inf a_n = L = \lim \sup a_n.$$

We want to show that $a_n \to L$. We will consider the case when $-\infty < L < \infty$, and leave the other cases ($L = \pm\infty$) to the reader as an exercise. There exists a positive integer N such that

$$n \ge N \quad \text{implies} \quad -\infty < i_n, s_n < \infty,$$

where i_n and s_n are the infimum and the supremum of the set $\{a_k : k \geq n\}$, respectively. Since

$$-\infty < i_n \leq a_n \leq s_n < \infty$$

for all $n \geq N$, it follows from the Squeeze Theorem 2.5 and $\lim i_n = L = \lim s_n$ that $\lim a_n = L$. $\qquad\qquad\square$

2.4.5   *Exercises*

(1) Find the limit inferior and the limit superior of each of the following sequences.

  (a) $a_n = n$ if n is odd, and $a_n = \frac{1}{n}$ if n is even.
  (b) $a_n = n - \frac{1}{n}$.
  (c) $a_n = (-1)^n n$.
  (d) $a_n = (-1)^n \frac{2n^2 - 3}{5n^2 + n - 1}$.
  (e) $a_n = \sqrt{n^2 - 2n} - n$.

(2) Finish the proof of Theorem 2.13 by showing that the limit inferior of any sequence $\{a_n\}$ is a subsequential limit.

(3) Finish the proof of Theorem 2.12 by showing that $\liminf a_n \leq L$ for any subsequential limit L of $\{a_n\}$.

(4) Finish the proof of Theorem 2.14 by showing that, if $\liminf a_n = \limsup a_n = L \in \{\pm\infty\}$, then $a_n \to L$.

(5) Prove that

$$\limsup(c + a_n) = c + \limsup a_n$$

and

$$\liminf(c + a_n) = c + \liminf a_n$$

for any sequence $\{a_n\}$ and real number c. Here we interpret $c + \infty$ as ∞ and $c - \infty$ as $-\infty$.

(6) Prove that $\liminf a_n \leq \limsup a_n$ for any sequence $\{a_n\}$.

(7) Prove the inequalities

$$\liminf a_n + \liminf b_n \leq \liminf(a_n + b_n)$$
$$\leq \limsup(a_n + b_n)$$
$$\leq \limsup a_n + \limsup b_n.$$

Give an example in which both the first and the third inequalities are strict.

(8) Let $\{a_{n_k}\}$ be a subsequence of $\{a_n\}$. Prove the inequalities

$$\liminf a_n \leq \liminf a_{n_k} \leq \limsup a_{n_k} \leq \limsup a_n.$$

(9) Let $\{a_n\}$ be a Cauchy sequence. Prove directly (i.e., without using Theorem 2.10) that $\liminf a_n = \limsup a_n$.

(10) Suppose that $a_n \leq b_n$ for every n.

(a) Prove that $\liminf a_n \leq \liminf b_n$.

(b) Prove that $\limsup a_n \leq \limsup b_n$.

(11) Let $\{a_n\}$ be a sequence, and let $\{l_n\}$ be a sequence in which each $l_n \in \mathbb{R}$ is a subsequential limit of $\{a_n\}$. Suppose that $l_n \to L \in \mathbb{R}$. Prove that L is a subsequential limit of $\{a_n\}$.

(12) Prove that $\limsup a_n = -\liminf(-a_n)$ for any sequence $\{a_n\}$.

(13) Prove that a sequence $\{a_n\}$ is convergent if and only if it is bounded and has exactly one subsequential limit.

(14) Prove that $a_n \to 0$ if and only if $\limsup|a_n| = 0$.

(15) Prove that a sequence $\{a_n\}$ is bounded if and only if $\limsup|a_n| < \infty$.

(16) Let $\{a_n\}$ be a bounded sequence.

   (a) Suppose that $\limsup a_n < M$ for some real number M. Prove that there exists a positive integer N such that $n \geq N$ implies $a_n < M$.

   (b) Suppose that $\liminf a_n > m$ for some real number m. Prove that there exists a positive integer N such that $n \geq N$ implies $a_n > m$.

(17) Let $\{a_n\}$ be a sequence.

   (a) Suppose that $\limsup a_n \in \mathbb{R}$. Prove that for every $\epsilon > 0$, there exists a positive integer N such that $n \geq N$ implies $a_n < (\limsup a_n) + \epsilon$.

   (b) Suppose that $\liminf a_n \in \mathbb{R}$. Prove that for every $\epsilon > 0$, there exists a positive integer N such that $n \geq N$ implies $a_n > (\liminf a_n) - \epsilon$.

(18) Let $\{a_n\}$ be a bounded sequence, and let r be a positive real number.

   (a) Prove that $\limsup(ra_n) = r\,(\limsup a_n)$.

   (b) Prove that $\liminf(ra_n) = r\,(\liminf a_n)$.

(19) Suppose that $a_n \to L$, a positive real number, and that $\{b_n\}$ is any sequence. Prove that $\limsup(a_n b_n) = L\,(\limsup b_n)$.

(20) Let $\{a_n\}$ and $\{b_n\}$ be bounded sequences of non-negative real numbers. Prove that

$$\limsup(a_n b_n) \leq (\limsup a_n)(\limsup b_n).$$

Give an example in which this inequality is strict.

(21) Let $\{a_n\}$ be a sequence. Prove that

$$\limsup a_n = \inf\{s_n : n \in \mathbb{Z}_+\} \quad \text{and} \quad \liminf a_n = \sup\{i_n : n \in \mathbb{Z}_+\},$$

where $s_n = \sup\{a_k : k \geq n\}$ and $i_n = \inf\{a_k : k \geq n\}$.

2.5 Additional Exercises

(1) Let r be an arbitrary real number.

   (a) Prove that there exists a strictly increasing sequence of rational numbers $\{a_n\}$ such that $a_n \to r$.

(b) Prove that there exists a strictly increasing sequence of irrational numbers $\{a_n\}$ such that $a_n \to r$.

(c) Repeat the above two parts with strictly decreasing instead of strictly increasing.

(2) Let $a_1 = \sqrt{2}$ and $a_{n+1} = \sqrt{2 + a_n}$ for $n \geq 1$.

(a) Prove that $a_n < 2$ for all n.

(b) Prove that $a_n < a_{n+1}$ for all n.

(c) Conclude that the sequence $\{a_n\}$ is convergent.

(3) Prove directly from the definition that the following sequences are Cauchy sequences.

(a) $a_n = \frac{2}{1+n^2}$.

(b) $a_n = \sum_{k=1}^{n} \frac{k^3}{3^k}$.

(c) $a_n = \sum_{k=1}^{n} \frac{2^k}{k^2 5^k}$.

(4) Prove that the sequence with

$$a_n = 1 + \frac{1}{\sqrt{2}} + \frac{1}{\sqrt{3}} + \cdots + \frac{1}{\sqrt{n}}$$

is divergent.

(5) Let p be a real number with $0 < p < 1$. Prove that the sequence with

$$a_n = 1 + \frac{1}{2^p} + \frac{1}{3^p} + \cdots + \frac{1}{n^p}$$

is divergent.

(6) If $\lim a_{2n-1} = L = \lim a_{2n}$, prove that $\lim a_n = L$.

(7) Suppose that $a_n \to L \neq 0$ and that $\{a_n b_n\}$ is a convergent sequence. Prove that $\{b_n\}$ is convergent.

(8) Suppose that $a_n \to L$ and that $f: \mathbb{Z}_+ \to \mathbb{Z}_+$ is a bijection. Consider the sequence with $b_n = a_{f(n)}$. Prove that $b_n \to L$.

(9) Let $p(x) = c_m x^m + c_{m-1} x^{m-1} + \cdots + c_0$ be a polynomial, so the c_i are real numbers, and suppose that $a_n \to L$. Prove that $p(a_n) \to p(L)$.

(10) Suppose that each $a_n > 0$ such that $\lim \left(\frac{a_{n+1}}{a_n} \right) = L$.

(a) If $L < 1$, prove that there exist a positive real number r with $L < r < 1$ and a positive integer N such that $a_{N+k} < r^k a_N$ for all positive integers k.

(b) If $L < 1$, use the previous part to prove that $a_n \to 0$.

(c) If $L > 1$, prove that $a_n \to \infty$.

(d) Given an example in which $L = 1$ and $\{a_n\}$ is convergent.

(e) Given an example in which $L = 1$ and $\{a_n\}$ is divergent.

(11) Let $p(x) = c_m x^m + c_{m-1} x^{m-1} + \cdots + c_0$ and $q(x) = d_k x^k + d_{k-1} x^{k-1} + \cdots + d_0$ be polynomials with $d_k \neq 0$. Prove that

$$\lim \frac{p(n)}{q(n)} = \begin{cases} \dfrac{c_m}{d_k} & \text{if } m = k, \\ 0 & \text{if } m < k, \\ \pm\infty & \text{if } m > k. \end{cases}$$

(12) Let $\{a_n\}$ be a sequence. Define another sequence by
$$b_n = \frac{a_1 + \cdots + a_n}{n}.$$

    (a) If $a_n \to L$, prove that $b_n \to L$.

    (b) Given an example in which $\{b_n\}$ is convergent but $\{a_n\}$ is divergent.

(13) Consider the sequence with $a_n = \left(1 + \frac{1}{n}\right)^n$.

    (a) Using Theorem 1.7 prove that
$$a_n = 1 + 1 + \frac{1}{2!}\left(1 - \frac{1}{n}\right) + \frac{1}{3!}\left(1 - \frac{1}{n}\right)\left(1 - \frac{2}{n}\right) + \cdots$$
$$+ \frac{1}{n!}\left(1 - \frac{1}{n}\right)\left(1 - \frac{2}{n}\right)\cdots\left(1 - \frac{n-1}{n}\right).$$

    (b) Use the previous part to show that the sequence $\{a_n\}$ is increasing.

    (c) Prove that
$$a_n \le 1 + 1 + \frac{1}{2!} + \cdots + \frac{1}{n!} < 3$$

    for each $n \ge 2$.

    (d) Conclude that $\{a_n\}$ is a convergent sequence.

(14) Find the limit inferior and limit superior of the sequence whose nth term is $3 + (-1)^n \left(2 - \frac{5}{n}\right)$. Does this sequence converge?

(15) Consider two sequences $\{a_n\}$ and $\{b_n\}$ that are equal as sets. Does it follow that these two sequences have the same set of subsequential limits?

(16) Construct a sequence whose set of subsequential limits is $\{0\} \cup \{\frac{1}{n} : n \in \mathbb{Z}_+\}$.

(17) Prove that there exists a sequence $\{a_n\}$ whose set of subsequential limits is exactly the closed interval $[0, 1]$.

(18) Let $a < b$ be real numbers.

    (a) Prove that there exists a sequence $\{r_n\}$ in (a, b) such that every rational number in (a, b) occurs in this sequence exactly once.

    (b) Find the set of subsequence limits of the sequence $\{r_n\}$.

(19) Let $\{a_n\}$ be a bounded sequence.

    (a) Prove that $\limsup = L$ if and only if for every $\epsilon > 0$, the inequality $|a_n - L| < \epsilon$ holds for an infinite number of a_n, and only finitely many $a_n > L + \epsilon$.

    (b) Prove that $\liminf = L$ if and only if for every $\epsilon > 0$, the inequality $|a_n - L| < \epsilon$ holds for an infinite number of a_n, and only finitely many $a_n < L - \epsilon$.

(20) Prove that the following statements are equivalent.

    (a) The Bolzano-Weierstrass Theorem 2.8.

    (b) The Nested Interval Property (Exercise (16) on page 44).

    (c) Every Cauchy sequence is convergent.

    (d) The Completeness Axiom.

In other words, prove that each statement implies the others. It suffices, for example, to prove the implications (a) $\Rightarrow$ (b) $\Rightarrow$ (c) $\Rightarrow$ (d) $\Rightarrow$ (a).

Chapter 3

Series

In this chapter we discuss a special kind of sequences, called *series*. In fact, we already discussed several examples of series in the previous chapter. Examples 2.11, 2.12, 2.18, and 2.19 are all series. In general, a series $\{s_n\}$ has the form $s_n = a_1 + \cdots + a_n$. As with sequences, the main question about a series is whether it converges.

In section 3.1 we discuss two convergence criteria for series due to Cauchy. The first one is the Cauchy Convergence Criterion (Theorem 2.10) for sequences, stated in the context of series. The second one is called the Cauchy Condensation Test. The two most basic types of series, the geometric series and the p-series, are also discussed. Several more tests for convergence of series are discussed in sections 3.2 and 3.3. They include the two Comparison Tests and the Alternating Series Test. The reader should already be familiar with these tests from calculus.

In section 3.4 we discuss the related concept of absolute convergence. The Ratio Test and the Root Test are discussed. In section 3.5 we consider what can happen when the terms of a series are rearranged. It turns out that for some convergent series, such a rearrangement may converge to a *different* limit.

To motivate the discussion of series, consider the decimal form of a real number

$$r = n.n_1 n_2 n_3 n_4 \ldots = n + 0.n_1 n_2 n_3 n_4 \ldots,$$

where n is an integer and each $n_k \in \{0, 1, \ldots, 9\}$. The kth digit n_k after the decimal point represents the real number $\frac{n_k}{10^k}$. So writing a real number r in its decimal form is actually representing it as a "sum"

$$r = n + \frac{n_1}{10} + \frac{n_2}{10^2} + \frac{n_3}{10^3} + \cdots. \tag{3.1}$$

However, this "sum" may go on indefinitely. For example, $\pi = 3.14159...$ has an unending decimal expansion. In other words, we are really considering the sequence $\{r_k\}$ in which

$$r_k = n + \frac{n_1}{10} + \frac{n_2}{10^2} + \cdots + \frac{n_k}{10^k},$$

and the limit of this sequence is r. As with any other sequence, here is the main question. Are we sure that this sequence converges? Can there be decimal expansions that do not converge? If so, which ones?

The example and questions above illustrate the need for a rigorous definition and careful analysis of series. With the tools that we will develop in this chapter, it is not hard to show that the sequence $\{r_k\}$ above converges. Of course, series are widely used in the mathematical sciences and are not restricted to decimal expansions.

3.1   Convergence of Series

In this section, we discuss some basic tests for convergence of series. We also discuss some examples of series that are useful for comparing with other series.

3.1.1   *Definition of a Series*

Definition 3.1. A **series** is a sequence $\{s_n\}$ in which s_n has the form

$$s_n = a_1 + a_2 + \cdots + a_n$$

for some sequence $\{a_n\}$. We also write such a series as

$$\sum_{n=1}^{\infty} a_n \quad \text{or} \quad \sum a_n,$$

and call s_n the nth **partial sum** of the series. A series $\sum a_n$ is said to be **convergent** if the sequence $\{s_n\}$ of partial sums is a convergent sequence. If L is the limit of $\{s_n\}$, we write

$$\sum a_n = L.$$

A series that is not convergent is called **divergent**. If $s_n \to \pm\infty$, then we say that the series **diverges to** $\pm\infty$ and write $\sum a_n = \pm\infty$.

As in the case of sequences, sometimes the a_n involved may not start with a_1 but some other term a_k. For example, we may have $a_n = (\log n)^{-1}$, which does not make sense if $n = 1$. In such cases, we start with $s_k = a_k$, and we have

$$s_n = a_k + a_{k+1} + \cdots + a_n$$

for $n \geq k$. We write

$$\sum_{n=k}^{\infty} a_n$$

if we wish to emphasize that the first term in the series is a_k. We will sometimes write

$$\sum a_n = a_1 + a_2 + a_3 + \cdots$$

just to exhibit the first few a_k in the series.

3.1.2 *Cauchy Convergence Criterion for Series*

Since a series $\sum a_n$ is actually a sequence $\{s_n\}$ of partial sums, any convergence criterion for sequences automatically applies to series. In particular, the Cauchy Convergence Criterion for sequences can be used for series as well.

Theorem 3.1. *A series $\sum a_n$ is convergent if and only if for every $\epsilon > 0$, there exists a positive integer N such that*

$$n > m \geq N \quad implies \quad \left| \sum_{k=m+1}^{n} a_k \right| < \epsilon.$$

Proof. By Theorem 2.10 the sequence $\{s_n\}$ of partial sums is convergent if and only if for every $\epsilon > 0$, there exists a positive integer N such that

$$n, m \geq N \quad implies \quad |s_n - s_m| < \epsilon.$$

If $n = m$ then this inequality is trivially true. If $n \neq m$, then either $n > m$ or $m > n$. In the case that $n > m$, we have

$$|s_n - s_m| = \left| \sum_{k=1}^{n} a_k - \sum_{k=1}^{m} a_k \right| = |a_{m+1} + a_{m+2} + \cdots + a_n| = \left| \sum_{k=m+1}^{n} a_k \right|.$$

In the case that $m > n$, we simply reverse the roles of n and m in the above discussion. □

Taking the negation of the Cauchy Convergence Criterion for series, we obtain a criterion for a series to diverge.

Corollary 3.1. *A series $\sum a_n$ is divergent if and only if there exists $\epsilon_0 > 0$ such that, for every positive integer N, there exist integers*

$$n > m \geq N \quad such\ that \quad \left| \sum_{k=m+1}^{n} a_k \right| \geq \epsilon_0.$$

Another useful criterion for showing that a series diverges is the following result.

Corollary 3.2. *If the series $\sum a_n$ is convergent, then $\lim a_n = 0$. Equivalently, if $\{a_n\}$ does not converge to 0, then the series $\sum a_n$ is divergent.*

Proof. Given $\epsilon > 0$, by Theorem 3.1 there exists a positive integer N such that

$$n > m \geq N \quad implies \quad \left| \sum_{k=m+1}^{n} a_k \right| < \epsilon.$$

In particular, with $n = m + 1 > N$, we have

$$|a_n - 0| = \left| \sum_{k=m+1}^{m+1} a_k \right| < \epsilon.$$

This shows that $a_n \to 0$. □

Let us emphasize that Corollary 3.2 above is a one-way implication; its converse is false. In other words, $\lim a_n = 0$ alone is *not* enough to guarantee the convergence of the series $\sum a_n$. This is illustrated in the following example.

3.1.3   *Harmonic and Geometric Series*

Example 3.1. The **harmonic series** is the series

$$\sum_{n=1}^{\infty} \frac{1}{n} = 1 + \frac{1}{2} + \frac{1}{3} + \cdots.$$

Since $a_n = \frac{1}{n}$, we have $a_n \to 0$ (Example 2.2). We now show that the harmonic series is divergent using Corollary 3.1 with $\epsilon_0 = \frac{1}{2}$. Given any positive integer N, we can take $m = N$ and $n = 2m = 2N$. We have

$$\sum_{k=N+1}^{2N} \frac{1}{k} = \underbrace{\frac{1}{N+1} + \frac{1}{N+2} + \cdots + \frac{1}{2N}}_{N \text{ terms}} \geq N \cdot \frac{1}{2N} = \frac{1}{2}.$$

The above inequality holds because each of the N terms is greater than or equal to $\frac{1}{2N}$. This shows that the harmonic series $\sum \frac{1}{n}$ is divergent, even though $\frac{1}{n} \to 0$.

Example 3.2. Let r be a real number. The **geometric series** is a series of the form

$$\sum_{n=0}^{\infty} r^n = 1 + r + r^2 + r^3 + \cdots.$$

If $|r| \geq 1$, then $|r|^n \geq 1$ for all n, and the sequence $\{r^n\}$ does not converge to 0. So by Corollary 3.2 the geometric series $\sum r^n$ diverges when $|r| \geq 1$. When $|r| < 1$, the nth partial sum is

$$s_n = 1 + r + \cdots + r^{n-1} + r^n.$$

So we have

$$r s_n = r + r^2 + \cdots + r^n + r^{n+1} = s_n + r^{n+1} - 1.$$

Solving for s_n in this equality, we have

$$s_n = \frac{1 - r^{n+1}}{1 - r} = \frac{1}{1-r} - \frac{1}{1-r} \cdot r^{n+1}.$$

From Example 2.5 we know that $r^{n+1} \to 0$ if $|r| < 1$. Thus, we conclude that the geometric series $\sum_{n=0}^{\infty} r^n$ converges to $\frac{1}{1-r}$ if $|r| < 1$.

The two examples above, the harmonic series and the geometric series, are actually closely related. We will make this clear after the following convergence test.

3.1.4   *Cauchy Condensation Test*

Theorem 3.2. *Suppose that each $a_n \geq 0$ and that the sequence $\{a_n\}$ is decreasing. Then the series $\sum_{n=1}^{\infty} a_n$ is convergent if and only if the series*

$$\sum_{k=0}^{\infty} 2^k a_{2^k} = a_1 + 2a_2 + 4a_4 + \cdots \tag{3.2}$$

is convergent.

Ideas of Proof. Both implications are proved using the Monotone Convergence Theorem. Therefore, in each case, we will establish that the sequence of partial sums is bounded and monotone.

Proof. Consider the two sequences of partial sums involved:

$$s_n = a_1 + a_2 + a_3 + a_4 + \cdots + a_n,$$
$$t_n = a_1 + 2a_2 + 4a_4 + 8a_8 \cdots + 2^n a_{2^n}$$
$$= a_1 + a_2 + a_2 + a_4 + \cdots + a_{2^n}.$$

It follows from the hypotheses on the a_k that

$$0 \le s_n \le t_n \tag{3.3}$$

for each $n \ge 1$. To prove the "if" part, suppose that the series (3.2) is convergent. In other words, we assume that the sequence $\{t_n\}$ of its partial sums is convergent. So the sequence $\{t_n\}$ is bounded (Theorem 2.2), which implies by (3.3) that the sequence $\{s_n\}$ is bounded. Since $a_k \ge 0$ for each k, the sequence $\{s_n\}$ is increasing and bounded. The Monotone Convergence Theorem 2.3 now implies that $\{s_n\}$, and hence the series $\sum a_n$, is convergent.

To prove the "only if" part, assume that $\sum a_n$, and hence $\{s_n\}$, is convergent. Since $\{s_n\}$ is a bounded sequence, there exists $M > 0$ such that

$$0 < s_n < M$$

for all n. Since the sequence $\{t_n\}$ of partial sums is increasing, to show that it is convergent, it is enough to show that it is bounded. If $n = 2^k$, since the sequence $\{a_n\}$ is decreasing, we have

$$2s_n = a_1 + a_1 + \underbrace{a_2 + a_2}_{\ge 2a_2} + \underbrace{a_3 + a_3 + a_4 + a_4}_{\ge 4a_4} + \cdots + a_n + a_n \ge a_1 + t_k \ge t_k.$$

We conclude that

$$t_k \le 2s_{2^k} < 2M$$

for all k. This shows that $\{t_n\}$ is bounded and monotone, and hence the series (3.2) is convergent. $\square$

The Cauchy Condensation Test is a remarkable result. It says that the convergence of the series $\sum a_n$ is determined by that of a series formed only with the terms a_{2^k} $(k \in \mathbb{N})$.

Example 3.3. Generalizing the harmonic series $\sum \frac{1}{n}$ (Example 3.1), we consider the *p*-**series**

$$\sum_{n=1}^{\infty} \frac{1}{n^p} = 1 + \frac{1}{2^p} + \frac{1}{3^p} + \cdots,$$

where p is a fixed real number. If $p \le 0$, then $n^{-p} \ge 1$ for all n, and $\{n^{-p}\}$ does not converge to 0. So by Corollary 3.2 the *p*-series $\sum n^{-p}$ is divergent when $p \le 0$.

Suppose that $p > 0$. In this case, the sequence $\{n^{-p}\}$ is decreasing and each $n^{-p} > 0$. Thus, by the Cauchy Condensation Test, the p-series is convergent if and only if the series

$$\sum_{n=0}^{\infty} 2^n \cdot \frac{1}{(2^n)^p} = \sum_{n=0}^{\infty} \left[2^{(1-p)}\right]^n$$

is convergent. This is a geometric series $\sum r^n$ with $r = 2^{(1-p)}$. By Example 3.2 this geometric series is convergent if and only if $2^{(1-p)} < 1$, i.e., $p > 1$. Therefore, we conclude that the p-series $\sum n^{-p}$ is convergent if and only if $p > 1$.

3.1.5 Exercises

(1) Let $\sum a_n = A$ and $\sum b_n = B$ be two convergent series, and let c be a real number.
   (a) Prove that $\sum(a_n + b_n) = A + B$.
   (b) Prove that $\sum c a_n = cA$.

(2) Let a and r be real numbers. Prove that the series $\sum_{n=0}^{\infty} ar^n$, also called a geometric series, is convergent with limit $\frac{a}{1-r}$ when $|r| < 1$ and is divergent when $|r| \geq 1$.

(3) Prove that the harmonic series $\sum \frac{1}{n}$ diverges to ∞.

(4) Determine whether the following series converge.
   (a) $\sum[\log n]^{-1}$.
   (b) $\sum[n \log n]^{-1}$.
   (c) $\sum[n(\log n)(\log \log n)]^{-1}$.
   (d) $\sum[n(\log n)^p]^{-1}$, where p is a fixed real number.

(5) Prove that the convergence of a series $\sum a_n$ is not affected by inserting or deleting finitely many terms. In particular, if $\sum_{n=1}^{\infty} a_n$ is convergent, then so is $\sum_{n=k}^{\infty} a_n$ for every positive integer k.

(6) Let $\sum a_n$ be a convergent series. For each m, let $b_m = \sum_{n=m}^{\infty} a_n$. Prove that $b_m \to 0$.

(7) Suppose that $a_n \to L \in \mathbb{R}$. Prove that the series $\sum_{n=1}^{\infty}(a_n - a_{n+1})$ converges to $a_1 - L$.

(8) Prove that $\sum_{n=1}^{\infty}[n(n+1)]^{-1}$ converges to 1.

(9) Suppose that $a_n \to \infty$. Prove that $\sum(a_{n+1} - a_n) = \infty$.

(10) Prove that each of the following series diverges to ∞.
   (a) $\sum(\sqrt{n+1} - \sqrt{n})$.
   (b) $\sum(\log(n+1) - \log n)$.

(11) Suppose that the series $\sum |a_n|$ is convergent.
   (a) Prove that $\sum a_n$ is convergent.
   (b) Give an example in which $\sum a_n$ is convergent but $\sum |a_n|$ is divergent.

(12) Suppose that each $a_n \geq 0$. Prove that the series $\sum a_n$ is convergent if and only if its sequence $\{s_n\}$ of partial sums is bounded.

(13) Prove that the series (3.1) for decimal expansion always converges.

(14) Suppose that each $a_n \geq 0$ and that $\sum a_n$ is convergent. Suppose that $\{a_{n_k}\}$ is a subsequence of $\{a_n\}$.

    (a) Prove that $\sum_{k=1}^{\infty} a_{n_k}$ is convergent.

    (b) Is $\sum a_{n_k}$ still convergent if we do not assume that $a_n \geq 0$ for all n?

(15) Suppose that $\sum a_n$ is a convergent series with limit S. Define

$$b_1 = a_1 + \cdots + a_{n_1},$$
$$b_2 = a_{n_1+1} + \cdots + a_{n_2},$$
$$b_3 = a_{n_2+1} + \cdots + a_{n_3},$$

etc., where $1 \leq n_1 < n_2 < n_3 < \cdots$. Prove that $\sum b_n$ converges to S. This says that in a convergent series, if parentheses are introduced to form a new series, then the resulting series converges to the same limit.

(16) Prove that the converse of the previous exercise is false. In other words, exhibit an example in which $\sum b_n$ is convergent but $\sum a_n$ is divergent.

(17) Suppose that each $a_n \geq 0$ and that $\sum a_n$ is a convergent series with limit S. Let $f: \mathbb{Z}_+ \to \mathbb{Z}_+$ be a bijection, and define $b_n = a_{f(n)}$. Prove that $\sum b_n = S$. This says that if the terms of a convergent series with non-negative terms are rearranged in any order, then the resulting series converges to the same limit.

3.2   Comparison Tests

In this section we discuss two Comparison Tests for convergence of series.

3.2.1   *Comparison Test*

To motivate the first comparison test, consider the series

$$\sum_{n=1}^{\infty} \frac{1}{2n^2 + 5} = \frac{1}{7} + \frac{1}{13} + \frac{1}{23} + \cdots.$$

Since $n^2 < 2n^2 + 5$, taking the reciprocals we have

$$0 < \frac{1}{2n^2 + 5} < \frac{1}{n^2}.$$

Since the bigger series $\sum n^{-2}$ is a convergent p-series (Example 3.3), it seems that the smaller series $\sum (2n^2 + 5)^{-1}$ should be convergent as well. This is, indeed, the case, as the following test shows.

Theorem 3.3 (Comparison Test). *Suppose that*

$$0 \leq a_n \leq b_n \quad \text{for all} \quad n.$$

(1) If $\sum b_n$ is convergent, then $\sum a_n$ is convergent.
(2) If $\sum a_n$ is divergent, then $\sum b_n$ is divergent.

Proof.   Consider the sequences of partial sums

$$s_n = a_1 + \cdots + a_n \quad \text{and} \quad S_n = b_1 + \cdots + b_n.$$

Both are increasing sequences, and

$$0 \le s_n \le S_n$$

for each n. If the series $\sum b_n$ is convergent, then by definition the sequence $\{S_n\}$ is convergent, and hence bounded (Theorem 2.2). This implies that the increasing sequence $\{s_n\}$ is bounded as well. By the Monotone Convergence Theorem 2.3, the sequence $\{s_n\}$, and hence the series $\sum a_n$, is convergent. This proves the first assertion. The other assertion is the contrapositive of the first assertion.   □

Example 3.4. Applying the Comparison Test to the series $\sum (2n^2 + 5)^{-1}$, we conclude from the convergence of the p-series $\sum n^{-2}$ and $0 < (2n^2 + 5)^{-1} < n^{-2}$ that $\sum (2n^2 + 5)^{-1}$ is convergent.

There are two common pitfalls when using the Comparison Test. First, the condition $0 \le a_n$ for all n, or at least all $n \ge N$ for some fixed N, cannot be omitted, as the following example illustrates.

Example 3.5. Consider the case when

$$a_n = -1 < n^{-2} = b_n.$$

Then $\sum b_n$ is convergent, but $\sum a_n$ is divergent. The Comparison Test does not apply here because the condition, $0 \le a_n$ for all n, is not satisfied.

Another common misconception of the Comparison Test is that the convergence of $\sum a_n$ implies that of $\sum b_n$. This is false, as the following example illustrates.

Example 3.6. Consider the case when

$$0 < a_n = n^{-2} \le 1 = b_n.$$

Then $\sum a_n$ is convergent, but $\sum b_n$ is divergent. The Comparison Test does not apply here because the hypotheses for both cases of the Comparison Test are not satisfied.

3.2.2   *Limit Comparison Test*

To motivate the second Comparison Test, consider the series

$$\sum_{n=1}^{\infty} \frac{2n}{5n^2 - 4} = 2 + \frac{4}{16} + \frac{6}{41} + \cdots.$$

When n is large, the quotient $\frac{2n}{5n^2-4}$ is roughly equal to $\frac{2}{5n}$. So the convergence behaviors of the two series $\sum \frac{2n}{5n^2-4}$ and $\sum \frac{2}{5n}$ should be the same. Since the harmonic series $\sum \frac{1}{n}$ is divergent (Example 3.1), so is $\sum \frac{2}{5n}$. Thus, we expect the series $\sum \frac{2n}{5n^2-4}$ to be divergent as well. This is true, as the following test shows.

Theorem 3.4 (Limit Comparison Test). *Suppose that*

$$a_n, b_n > 0 \quad \text{for all} \quad n \quad \text{and} \quad \lim_{n \to \infty} \frac{a_n}{b_n} = L \in \mathbb{R}.$$

(1) If $L > 0$, then $\sum a_n$ is convergent if and only if $\sum b_n$ is convergent.
(2) If $L = 0$ and $\sum b_n$ is convergent, then $\sum a_n$ is convergent.

Ideas of Proof. The plan is to show that we can actually use the Comparison Test (Theorem 3.3).

Proof. If $\lim \frac{a_n}{b_n} = L > 0$, then there exists a positive integer N such that

$$n \geq N \quad \text{implies} \quad \frac{L}{2} < \frac{a_n}{b_n} < \frac{3L}{2},$$

which is equivalent to

$$\frac{L}{2} b_n < a_n < \frac{3L}{2} b_n. \tag{3.4}$$

If $\sum a_n$ is convergent, then so is $\frac{2}{L} \sum a_n$. So the left inequality in (3.4) and the Comparison Test imply that $\sum b_n$ is convergent. Conversely, if $\sum b_n$ is convergent, then so is $\frac{3L}{2} \sum b_n$. So the right inequality in (3.4) and the Comparison Test imply that $\sum a_n$ is convergent. The proof of the other assertion is an exercise. □

Example 3.7. For the series $\sum_{n=1}^{\infty} a_n$ with $a_n = \frac{2n}{5n^2-4}$, we use the Limit Comparison Test with $b_n = \frac{1}{n}$. Then $\frac{a_n}{b_n} \to \frac{2}{5} > 0$, so the divergence of the harmonic series $\sum \frac{1}{n}$ implies the divergence of the series $\sum \frac{2n}{5n^2-4}$.

3.2.3 Exercises

(1) Prove Theorem 3.4 when $L = 0$.
(2) Test the following series for convergence.

  (a) $\sum (3n^2 - n + 4)^{-1}$.
  (b) $\sum (4n - 3)^{-1/2}$.
  (c) $\sum (2n + 3\sqrt{n})^{-1}$.
  (d) $\sum (2n + 3\sqrt{n})^{-2}$.
  (e) $\sum (n!)^{-1}$.
  (f) $\sum (n2^n)^{-1}$.

(3) Suppose that $a_n, b_n > 0$ for each n and that $\frac{a_n}{b_n} \to \infty$. If $\sum b_n$ is divergent, prove that $\sum a_n$ is divergent.
(4) If $\sum a_n$ is convergent and each $a_n \geq 0$, prove that $\sum a_n^2$ is convergent.
(5) Let $p(x) = \sum_{i=0}^{m} c_i x^i$ and $q(x) = \sum_{j=0}^{k} d_j x^j$ be polynomials with $c_m, d_k \neq 0$. Prove that the series $\sum \frac{p(n)}{q(n)}$ is convergent if and only if $k \geq m + 2$. Here we assume that the series starts at a term $\frac{p(N)}{q(N)}$ with N sufficiently large so that $q(m) \neq 0$ for all $m \geq N$. This is known as the **Polynomial Test**.

(6) Suppose that $a_n, b_n \geq 0$ for all n.

    (a) Prove that

$$\sqrt{a_n b_n} \leq \frac{a_n + b_n}{2}$$

    for all n.

    (b) Use the previous part to prove that $\sum \sqrt{a_n b_n}$ is convergent if both $\sum a_n$ and $\sum b_n$ are convergent.

3.3   Alternating Series Test

In this section we discuss a convergence test for series whose terms have alternating signs.

Definition 3.2. An **alternating series** is a series $\sum(-1)^{n+1} a_n$ or $\sum(-1)^n a_n$ with each $a_n > 0$.

Example 3.8. The **alternating harmonic series**

$$\sum_{n=1}^{\infty} (-1)^{n+1} \frac{1}{n} = 1 - \frac{1}{2} + \frac{1}{3} - \frac{1}{4} + \cdots$$

considered in Example 2.18 is an alternating series. There we showed that this series is convergent, in strong contrast with the harmonic series $\sum \frac{1}{n}$ (Example 3.1). The crucial ingredient in Example 2.18 is that the sequence $\{\frac{1}{n}\}$ is decreasing. Of course, it is also true that $\frac{1}{n} \to 0$. This example illustrates the ideas behind the following test.

Theorem 3.5 (Leibnitz Alternating Series Test). *Suppose that*

- *$\{a_n\}$ is decreasing with each $a_n > 0$, and*
- *$\lim a_n = 0$.*

Then the alternating series $\sum(-1)^{n+1} a_n$ is convergent.

Ideas of Proof. The plan is to recycle the argument in Example 2.18. We will show that the alternating series $\sum(-1)^{n+1} a_n$ satisfies the Cauchy Convergence Criterion (Theorem 3.1).

Proof. Let $\epsilon > 0$ be given. Since $a_n \to 0$, there exists a positive integer N such that

$$m \geq N \quad \text{implies} \quad a_m < \epsilon.$$

Then for $n > m \geq N$ we have

$$\left| \sum_{k=m+1}^{n} (-1)^{k+1} a_k \right| = \left| (-1)^{m+2} a_{m+1} + (-1)^{m+3} a_{m+2} + \cdots + (-1)^{n+1} a_n \right|$$

$$= \left| a_{m+1} - a_{m+2} + a_{m+3} - \cdots + (-1)^{m+n+1} a_n \right|$$

$$= \left| a_{m+1} - \underbrace{(a_{m+2} - a_{m+3})}_{\text{non-negative}} - \underbrace{(a_{m+4} - a_{m+5})}_{\text{non-negative}} - \cdots \right|$$

$$\leq a_{m+1} < \epsilon.$$

This shows that the series $\sum (-1)^{n+1} a_n$ satisfies the Cauchy Convergence Criterion and is, therefore, convergent.   □

Example 3.9. Consider the alternating series

$$\sum (-1)^{n+1} \frac{2^{\frac{1}{n}}}{n}.$$

We have that $n^{-1} 2^{\frac{1}{n}} > 0$ and that the sequence $\{n^{-1} 2^{\frac{1}{n}}\}$ is decreasing. Moreover, we have $n^{-1} \to 0$ and $1 < 2^{\frac{1}{n}} \leq 2$ for all n, so $n^{-1} 2^{\frac{1}{n}} \to 0$. Thus, the hypotheses of the Alternating Series Test are all satisfied, and we conclude that the alternating series is convergent.

One common mistake that students make about the Alternating Series Test is to use it to show that an alternating series is divergent. The Alternating Series Test *cannot* be used to show that any series is divergent. If one of its hypotheses is not satisfied, then the test simply does not apply. In this case, the series may converge or diverge.

Example 3.10. The alternating series

$$\sum (-1)^n \frac{2n}{3n - 1}$$

is divergent because $\frac{2n}{3n-1} \to \frac{2}{3}$, and so $\{(-1)^n \frac{2n}{3n-1}\}$ does not converge to 0 (Corollary 3.2). The Alternating Series Test does not apply to this alternating series.

The Alternating Series Test also does not apply to the alternating series

$$\sum (-1)^{n+1} a_n = 1 - \frac{1}{3^2} + \frac{1}{2^2} - \frac{1}{5^2} + \frac{1}{4^2} - \cdots$$

because $\{a_n\}$ is not decreasing. However, this alternating series is convergent by Exercises (11) and (17) on page 66 and Example 2.19.

3.3.1 *Exercises*

(1) Test the following series for convergence.

    (a) $\sum (-1)^{n+1} (\log n)^{-1}$.

    (b) $\sum (-1)^{n+1} n^2 (n!)^{-1}$.

    (c) $\sum (-1)^{n+1} n^{-2}$.

(d) $\sum(-1)^{n+1}n^3 3^{-n}$.

(2) Prove or disprove: If $\sum a_n$ is a convergent series and if $b_n \to L$, then $\sum a_n b_n$ is convergent.

(3) Prove or disprove: If $\sum a_n$ is convergent, then $\sum a_n^2$ is convergent.

(4) Suppose that $a_n, b_n \geq 0$ for all n and that $\sum a_n$ and $\sum b_n$ are both convergent.

    (a) Prove that $\sum a_n b_n$ is convergent.

    (b) Is $\sum a_n b_n$ still convergent if the hypothesis $a_n, b_n \geq 0$ for all n is omitted?

(5) Suppose that $\sum(-1)^{n+1}a_n$ is an alternating series that satisfies the hypotheses of the Alternating Series Test and has limit L. Prove that $0 < L < a_1$.

(6) Consider the alternating series $\sum(-1)^{n+1}a_n$ in which $a_n = \frac{1}{n}$ if n is odd and $a_n = \frac{1}{n^2}$ if n is even.

    (a) Prove that $a_n \to 0$.

    (b) Prove that $a_n - a_{n+1} > \frac{1}{n+1}$ for all odd n.

    (c) Prove that the alternating series $\sum(-1)^{n+1}a_n$ is divergent.

This series illustrates that the decreasing hypothesis in the Alternating Series Test cannot be omitted.

3.4   Absolute Convergence

Recall that the alternating harmonic series $\sum(-1)^{n+1}\frac{1}{n}$ is convergent (Example 2.18), but the harmonic series $\sum \frac{1}{n}$ is divergent (Example 3.1). These two series tell us that, in general, $\sum a_n$ and $\sum |a_n|$ can have very different convergence behaviors. To make precise the relationships between such series, we make the following definitions.

3.4.1   *Absolute and Conditional Convergence*

Definition 3.3. A series $\sum a_n$ is said to be

- **absolutely convergent** if the series $\sum |a_n|$ is convergent.
- **conditionally convergent** if it is convergent, but $\sum |a_n|$ is divergent.

The alternating harmonic series $\sum(-1)^{n+1}\frac{1}{n}$ is an example of a conditionally convergent series. So convergence does *not* imply absolute convergence. However, the other implication is true.

Theorem 3.6. *If a series $\sum a_n$ is absolutely convergent, then it is convergent.*

Proof.   We will show that the series $\sum a_n$ satisfies the Cauchy Convergence Criterion (Theorem 3.1). Since $\sum a_n$ is absolutely convergent, given $\epsilon > 0$ there exists a positive integer N such that $n > m \geq N$ implies

$$\left| \sum_{k=m+1}^{n} a_k \right| \leq \sum_{k=m+1}^{n} |a_k| = \left| \sum_{k=m+1}^{n} |a_k| \right| < \epsilon.$$

This shows that $\sum a_n$ is convergent.   □

In other words, *absolute convergence implies convergence, but not vice versa.* The following example illustrates that sometimes it is easier to show that a series is absolutely convergent than to show directly that it is convergent.

Example 3.11. Consider the series $\sum \frac{\cos(n)}{n^2}$. Some of its terms are negative because $\cos(n)$ can be negative for certain values of n. One would like to compare this series to the convergent p-series $\sum \frac{1}{n^2}$ with $p = 2$ (Example 3.3). However, neither one of the Comparison Tests applies because we cannot be sure that the nth term $\frac{\cos(n)}{n^2}$ is positive. By Theorem 3.6 to show that $\sum \frac{\cos(n)}{n^2}$ is convergent, it suffices to show that it is absolutely convergent. Since $|\cos(n)| \leq 1$ for any n, it follows that

$$0 \leq \left| \frac{\cos(n)}{n^2} \right| \leq \frac{1}{n^2}.$$

Now the Comparison Test applies. Since the p-series $\sum \frac{1}{n^2}$ is convergent, so is $\sum |\frac{\cos(n)}{n^2}|$. In other words, $\sum \frac{\cos(n)}{n^2}$ is absolutely convergent, and hence convergent.

The above example shows the usefulness of the concept of absolute convergence. Even if one is only interested in convergence, it is sometimes more convenient to consider absolute convergence.

3.4.2   Root Test

We now develop two commonly used tests for absolute convergence. Recall the concept of limit superior in section 2.4.

Theorem 3.7 (Cauchy Root Test). *Suppose that*

$$L = \limsup |a_n|^{\frac{1}{n}}.$$

Then the series $\sum a_n$ is

- *absolutely convergent if $L < 1$, and*
- *divergent if $L > 1$.*

Ideas of Proof. For large n, $|a_n|^{\frac{1}{n}}$ is roughly equal to L, so $|a_n|$ is approximately L^n. Thus, the series $\sum |a_n|$ behaves like the geometric series $\sum L^n$, which is convergent if and only if $|L| = L < 1$.

Proof. First suppose that $L < 1$. We need to show that $\sum |a_n|$ is convergent. We can choose an $\epsilon > 0$ such that

$$r = L + \epsilon < 1.$$

For example, the choice $\epsilon = \frac{1-L}{2}$ will do. Since L is the largest subsequential limit of the sequence $\{|a_n|^{\frac{1}{n}}\}$ (Theorems 2.12 and 2.13), this sequence is bounded. There are at most finitely many n such that

$$|a_n|^{\frac{1}{n}} > L + \epsilon.$$

Thus, we can choose a positive integer N such that

$$n \geq N \quad \text{implies} \quad |a_n|^{\frac{1}{n}} \leq r < 1,$$

which in turn implies

$$|a_n| \leq r^n. \tag{3.5}$$

Since $0 < r < 1$, the geometric series $\sum r^n$ is convergent. By (3.5) and the Comparison Test, the series $\sum_{n=N}^{\infty} |a_n|$, and hence $\sum_{n=1}^{\infty} |a_n|$, is convergent.

Next suppose that $L > 1$. We can choose an $\epsilon > 0$ such that $L - \epsilon > 1$. There are infinitely many n such that

$$|a_n|^{\frac{1}{n}} > L - \epsilon > 1,$$

which implies that $|a_n| > 1$ for infinitely many n. So $\{a_n\}$ does not converge to 0, and the series $\sum a_n$ is divergent (Corollary 3.2).   □

Example 3.12. Consider the series $\sum a_n$ with $a_n = \left(\frac{3n}{5n-2}\right)^n$. Then

$$0 < a_n^{\frac{1}{n}} = \frac{3n}{5n-2} \to \frac{3}{5}.$$

So in this case

$$\limsup |a_n|^{\frac{1}{n}} = \frac{3}{5} < 1$$

by Theorem 2.14. Thus, the Root Test says that the series $\sum a_n$ is absolutely convergent, and hence convergent.

The Root Test gives no information if $\limsup |a_n|^{\frac{1}{n}} = 1$. In other words, when $\limsup |a_n|^{\frac{1}{n}} = 1$, the series may converge or diverge (Exercise (3) below).

3.4.3   *Ratio Test*

The following test for absolute convergence will be used again when we discuss power series later in this book.

Theorem 3.8 (d'Alembert Ratio Test). *Suppose that $a_n \neq 0$ for all n.*

(1) If

$$\limsup \left| \frac{a_{n+1}}{a_n} \right| < 1,$$

then $\sum a_n$ is absolutely convergent.

(2) If

$$\liminf \left| \frac{a_{n+1}}{a_n} \right| > 1,$$

then $\sum a_n$ is divergent.

Ideas of Proof. Similar to the proof of the Root Test, the plan is to relate the series to a suitable geometric series, whose convergence behavior is known.

Proof. Suppose that

$$L = \limsup \left| \frac{a_{n+1}}{a_n} \right| < 1.$$

To show that $\sum |a_n|$ is convergent, first choose an $\epsilon > 0$ such that

$$r = L + \epsilon < 1.$$

There are at most finitely many n such that

$$\left| \frac{a_{n+1}}{a_n} \right| > r.$$

Thus, there exists a positive integer N such that

$$n \geq N \quad \text{implies} \quad \left| \frac{a_{n+1}}{a_n} \right| \leq r.$$

In particular, we have

$$|a_{N+1}| \leq r|a_N|, \quad |a_{N+2}| \leq r|a_{N+1}| \leq r^2 |a_N|,$$

and so forth, leading to

$$|a_{N+k}| \leq r^k |a_N| \quad \text{for all} \quad k \geq 0.$$

Therefore, we have

$$\sum_{l=1}^{k} |a_{N+l}| \leq \sum_{l=1}^{k} r^l |a_N| < \frac{|a_N|}{1-r}$$

for all $k \geq 1$, since $0 < r < 1$. This shows that the sequence of partial sums of the series $\sum_{l=1}^{\infty} |a_{N+l}|$, which is increasing, is bounded as well. So $\sum_{l=1}^{\infty} |a_{N+l}|$ is convergent by the Monotone Convergence Theorem 2.3. This implies that $\sum |a_n|$ is convergent.

The other assertion is left as an exercise for the reader. $\qquad\square$

Example 3.13. Consider the series $\sum a_n$ with $a_n = \frac{n}{2^n}$. We have

$$\left| \frac{a_{n+1}}{a_n} \right| = \left| \frac{n+1}{2^{n+1}} \cdot \frac{2^n}{n} \right| = \frac{n+1}{2n} \to \frac{1}{2} < 1.$$

By the Ratio Test the series $\sum a_n$ is absolutely convergent, and hence convergent.

Example 3.14. Consider the series $\sum a_n$ with $a_n = \frac{x^n}{n!}$, where x is any real number. Then

$$\left| \frac{a_{n+1}}{a_n} \right| = \left| \frac{x^{n+1}}{(n+1)!} \cdot \frac{n!}{x^n} \right| = \frac{|x|}{n+1} \to 0 < 1,$$

regardless of what x is. Therefore, by the Ratio Test the series $\sum \frac{x^n}{n!}$ is absolutely convergent, and hence convergent, for all real numbers x.

3.4.4　Exercises

(1) Test for convergence for the following series.

　(a) $\sum \frac{n^3}{2^n}$.

　(b) $\sum \frac{\sin n}{n^2}$.

　(c) $\sum (n!)^{-\frac{1}{2}}$.

　(d) $\sum (\log n)^{-n}$, $n \geq 2$.

(2) In each case, determine the real numbers x such that the series converges.

　(a) $\sum (-1)^n \frac{x^{2n}}{(2n)!}$.

　(b) $\sum (-1)^n \frac{x^{2n+1}}{(2n+1)!}$.

　(c) $\sum (-1)^{n+1} n x^n$.

　(d) $\sum \frac{x^n}{n}$.

　(e) $\sum n! x^n$.

(3) Denote $\limsup |a_n|^{\frac{1}{n}}$ by L.

　(a) Give an example of a convergent series $\sum a_n$ with $L = 1$.

　(b) Give an example of a divergent series $\sum a_n$ with $L = 1$.

　Such examples illustrate that the Root Test is inconclusive when $L = 1$.

(4) Finish the proof of Theorem 3.8. In other words, prove that if $a_n \neq 0$ for all n and $\liminf |\frac{a_{n+1}}{a_n}| > 1$, then $\sum a_n$ is divergent.

(5) Give examples to show that the Ratio Test is inconclusive when

$$\liminf \left| \frac{a_{n+1}}{a_n} \right| \leq 1 \leq \limsup \left| \frac{a_{n+1}}{a_n} \right|.$$

(6) Suppose that $a_n > 0$ for all n.

　(a) Let $L = \limsup \frac{a_{n+1}}{a_n}$ and $\epsilon > 0$. Prove that there exists a positive integer N such that

$$a_n \leq (L + \epsilon)^n \frac{a_N}{(L + \epsilon)^N}$$

　for all $n \geq N$.

　(b) Use the previous part to prove that

$$\limsup \left(a_n^{\frac{1}{n}} \right) \leq \limsup \left(\frac{a_{n+1}}{a_n} \right).$$

　(c) Prove that

$$\liminf \left(\frac{a_{n+1}}{a_n} \right) \leq \liminf \left(a_n^{\frac{1}{n}} \right).$$

　This exercise illustrates that when the Ratio Test is applicable, then so is the Root Test.

(7) Consider the series

$$\frac{1}{2} + \left(\frac{1}{2}\right)\left(\frac{3}{2}\right) + \left(\frac{1}{2}\right)^2 \left(\frac{3}{2}\right) + \left(\frac{1}{2}\right)^2 \left(\frac{3}{2}\right)^2 + \left(\frac{1}{2}\right)^3 \left(\frac{3}{2}\right)^2 + \cdots.$$

(a) Prove that $\liminf \frac{a_{n+1}}{a_n} = \frac{1}{2}$ and $\limsup \frac{a_{n+1}}{a_n} = \frac{3}{2}$. Thus, the Ratio Test is not applicable.

(b) Prove that $a_n^{\frac{1}{n}} \to \frac{\sqrt{3}}{2}$. Thus, the Root Test shows that $\sum a_n$ is convergent.

This example together with the previous exercise illustrate that the Root Test is more general than the Ratio Test.

(8) Suppose that $a_n > 0$ for all n and that $\limsup a_n^{\frac{1}{n}} < 1$. Prove that $\sum n^p a_n$ is convergent for all positive integers p.

(9) Using Exercise (6b) or otherwise, prove that $(n!)^{\frac{1}{n}} \to \infty$.

(10) Suppose that $\sum a_n^2$ and $\sum b_n^2$ are both convergent.

(a) Prove that $|a_n b_n| \le \frac{a_n^2 + b_n^2}{2}$ for each n.

(b) Prove that $\sum a_n b_n$ is absolutely convergent.

(11) Suppose that $a_n \ge 0$ for all n and that $\sum a_n$ is convergent. Using the previous exercise, prove that $\sum \frac{\sqrt{a_n}}{n}$ is convergent.

(12) Suppose that

$$|a_n| \le \frac{1}{n} - \frac{1}{n+1}$$

for all n. Prove that $\sum a_n$ is absolutely convergent.

(13) Consider the series

$$\sum a_n = 1 + \frac{1}{2} + \frac{1}{3} + \frac{1}{2^2} + \frac{1}{3^2} + \cdots.$$

(a) Prove that the Ratio Test is *not* applicable.

(b) Use the Root Test to determine if $\sum a_n$ is convergent.

(14) Suppose that $a_n > 0$ for all n and that $\lim \frac{a_{n+1}}{a_n} = L > 0$.

(a) Given $\epsilon > 0$ with $\epsilon < L$, prove that there exists a positive integer N such that

$$(L - \epsilon)^n \frac{a_N}{(L - \epsilon)^N} < a_n < (L + \epsilon)^n \frac{a_N}{(L + \epsilon)^N}$$

for all integers $n > N$.

(b) Use the previous part to prove that $a_n^{\frac{1}{n}} \to L$.

3.5   Rearrangement of Series

Suppose that $\sum a_n$ is a convergent series. If the terms a_n are rearranged in the series, does the resulting series converge to the same limit? We will discuss this question in this section.

Definition 3.4. If $f : \mathbb{Z}_+ \to \mathbb{Z}_+$ is a bijection, then

$$\sum a_{f(n)} = a_{f(1)} + a_{f(2)} + \cdots$$

is called a **rearrangement** of the series $\sum a_n$.

For a convergent sequence $\{a_n\}$, any rearrangement results in a sequence $\{a_{f(n)}\}$ that also converges to the same limit (Exercise (8) on page 58). First we observe that this has an analog for *absolutely* convergent series.

Theorem 3.9. *Let $\sum a_n$ be an absolutely convergent series with $\sum |a_n| = L$, and let*

$$\sum b_n = \sum a_{f(n)}$$

be a rearrangement of $\sum a_n$. Then $\sum b_n$ is absolutely convergent with $\sum |b_n| = L$.

Ideas of Proof. The absolute convergence of $\sum b_n$ will be established using the Monotone Convergence Theorem.

Proof. First we show that $\sum |b_n|$ is convergent. Denote by s_n and S_n the nth partial sums of the series $\sum |a_n|$ and $\sum |b_n|$, respectively. Both $\{s_n\}$ and $\{S_n\}$ are increasing. We have

$$S_n = \sum_{k=1}^{n} |b_k| = \sum_{k=1}^{n} |a_{f(k)}| \leq s_M \leq L, \tag{3.6}$$

where

$$M = \max\{f(1), \ldots, f(n)\}.$$

So $\{S_n\}$ is both increasing and bounded. The Monotone Convergence Theorem 2.3 implies that $\{S_n\}$, and hence $\sum |b_n|$, is convergent. Denote its limit by L'. We must show that $L = L'$. Equation (3.6) implies that

$$L' = \lim S_n \leq L.$$

On other hand, the series $\sum a_n$ is a rearrangement of $\sum b_n$, since

$$a_n = b_{f^{-1}(n)}.$$

So the same argument as above shows that

$$L \leq L'.$$

Thus, we conclude that $L = L'$. □

Corollary 3.3. *If $\sum a_n$ is a convergent series with each $a_n \geq 0$, then any rearrangement $\sum a_{f(n)}$ of $\sum a_n$ also converges to the same limit.*

Proof. When $a_n \geq 0$ for all n, we have

$$a_n = |a_n|,$$

and convergence of the series $\sum a_n$ is equivalent to absolute convergence. So we can apply Theorem 3.9. □

The situation for *conditionally* convergent series is drastically different. Recall that a conditionally convergent series $\sum a_n$ is a convergent series that is not absolutely convergent. We now show that given a conditionally convergent series, *every* real number L is the limit of a rearrangement of the series.

Theorem 3.10 (Riemann Rearrangement Theorem). *Let*

- $\sum a_n$ *be a conditionally convergent series, and*
- L *be an arbitrary real number.*

Then there exists a rearrangement of $\sum a_n$ that converges to L.

Ideas of Proof. The plan is to rearrange the positive and negative terms of the series such that the resulting partial sums s_n oscillate above and below the intended limit L. The convergence of $\{s_n\}$ to L is guaranteed by the fact that the positive terms sequence and the negative terms sequence both converge to 0.

Proof. Let p_n and q_n be the nth positive term and the nth negative term in $\{a_k\}$, respectively. Since $\sum a_n$ is convergent, we have $\lim a_n = 0$, which implies that

$$\lim p_n = 0 = \lim q_n.$$

Moreover, the conditional convergence of $\sum a_n$ implies that

$$\sum p_n = \infty \quad \text{and} \quad \sum q_n = -\infty$$

by Exercise (2) on page 80.

Now we construct the desired rearrangement. Take *just* enough, possibly zero, positive terms $p_1, \ldots, p_{n_1}$ so that their sum exceeds L. In other words, we have

$$p_1 + \cdots + p_{n_1-1} \le L < p_1 + \cdots + p_{n_1} = s_{n_1}.$$

This is possible because $\sum p_n = \infty$. Now add *just* enough negative terms to the partial sum s_{n_1} so that the sum is less than L, so

$$s_{n_1+n_2} = s_{n_1} + q_1 + \cdots + q_{n_2} < L \le s_{n_1} + q_1 + \cdots + q_{n_2-1}.$$

This is possible because $\sum q_n = -\infty$. This process is now repeated ad infinitum. Add just enough positive terms $p_{n_1+1}, \ldots, p_{n_1+n_3}$ to $s_{n_1+n_2}$ such that the sum $s_{n_1+n_2+n_3}$ exceeds L. Then add just enough negative terms to $s_{n_1+n_2+n_3}$ such that the sum is less than L, and so forth.

Let $\sum a_{f(n)}$ be the rearrangement of $\sum a_n$ constructed in the previous paragraph, and let $\{s_n\}$ be its sequence of partial sums. To show that $\{s_n\}$ converges to L, let $\epsilon > 0$ be given. Since

$$\lim p_n = 0 = \lim q_n,$$

there exists a positive integer N_0 such that

$$n \ge N_0 \quad \text{implies} \quad p_n < \frac{\epsilon}{2} \quad \text{and} \quad |q_n| < \frac{\epsilon}{2}.$$

Let N be a sufficiently large integer such that $\{a_n\}_{n=1}^{N}$ includes $\{p_n\}_{n=1}^{N_0}$ and $\{q_n\}_{n=1}^{N_0}$. Then for $n \ge N$ we have

$$|s_n - L| \le \sup\{p_k, |q_k| : k \ge N_0\} \le \frac{\epsilon}{2} < \epsilon.$$

This proves that $\lim s_n = L$. $\qquad\square$

The proof above can be modified such that L can be taken to be the extended real numbers $\pm\infty$. Another modification of the proof above gives a rearrangement that is divergent but does not diverge to $\pm\infty$ (Exercise (3) below).

3.5.1   *Exercises*

(1) Suppose that $\sum a_n$ converges absolutely and that $\sum b_n$ is a rearrangement of $\sum a_n$. If $\sum a_n = L$, prove that $\sum b_n = L$.

(2) Let $\sum a_n$ be a conditionally convergent series. Suppose that p_n is the nth positive term in $\{a_k\}$ and that q_n is the nth negative term in $\{a_k\}$.

    (a) Prove that there are infinitely many p_n and infinitely many q_n.
    (b) Prove that $p_n \to 0$ and that $q_n \to 0$.
    (c) Prove that $\sum p_n = \infty$ and $\sum q_n = -\infty$.

(3) Suppose that $\sum a_n$ is a conditionally convergent series.

    (a) Prove that there exists a rearrangement of $\sum a_n$ that diverges to ∞.
    (b) Prove that there exists a rearrangement of $\sum a_n$ that diverges to $-\infty$.
    (c) Prove that there exists a rearrangement of $\sum a_n$ that is divergent but does not diverge to $\pm\infty$.

(4) Suppose that $a_n \geq 0$ for all n and that $\sum a_n = \infty$. Prove that $\sum a_{f(n)} = \infty$ for every rearrangement $\sum a_{f(n)}$ of $\sum a_n$.

(5) Let $\sum a_n$ be a conditionally convergent series. Suppose that $A < B$, where A and B are extended real numbers. Prove that there exists a rearrangement $\sum a_{f(n)}$ such that

$$\liminf s_n = A \quad \text{and} \quad \limsup s_n = B,$$

where $\{s_n\}$ is the sequence of partial sums of $\sum a_{f(n)}$.

3.6   Additional Exercises

(1) Suppose that $\{a_n\}$ is a decreasing sequence with $a_n > 0$ for all n. If $\sum a_n$ is convergent, prove that $na_n \to 0$. This is known as **Abel's Theorem**. Compare Abel's Theorem with Corollary 3.2.

(2) Prove that $\sum \frac{\log n}{n^2}$ is convergent.

(3) Prove that $\sum \left(\frac{x}{n}\right)^n$ is convergent for every real number x.

(4) Suppose that

$$|a_n| \leq b_n - b_{n+1}$$

for all n, where $\{b_n\}$ is decreasing with $b_n \to 0$. Prove that $\sum a_n$ is absolutely convergent.

(5) In each case, determine if the series is convergent or divergent.

    (a) $1 + \frac{1}{4} + \frac{1}{7} + \frac{1}{10} + \frac{1}{13} + \cdots$
    (b) $1 + \frac{1}{2} - \frac{1}{3} + \frac{1}{4} + \frac{1}{5} - \frac{1}{6} + \cdots$
    (c) $1 - \frac{1}{2} - \frac{1}{3} + \frac{1}{4} + \frac{1}{5} - \frac{1}{6} - \frac{1}{7} + \cdots$

(6) Suppose that $a_n \geq 0$ for all n

    (a) If $\sum a_n$ is convergent, prove that $\sum \frac{a_n}{1+a_n}$ is convergent.

(b) If $\sum a_n$ is divergent, prove that $\sum \frac{a_n}{1+a_n}$ is divergent.

(7) Suppose that $\sum a_n$ is absolutely convergent and that $\{b_n\}$ is bounded. Prove that $\sum a_n b_n$ is absolutely convergent.

(8) Prove or disprove: If $\sum a_n$ is convergent and if $\{b_n\}$ is bounded, then $\sum a_n b_n$ is convergent.

(9) Suppose that $a_n \to 0$. Define $b_n = a_{2n-1} + a_{2n}$. Suppose that $\sum b_n$ is convergent.

   (a) Prove that $\sum a_n$ is convergent.

   (b) Show by an example that $\sum a_n$ can be divergent if the hypothesis $a_n \to 0$ is omitted.

(10) Suppose that $a_n \to 0$. Prove that there exists a subsequence a_{n_k} such that $\sum_{k=1}^{\infty} a_{n_k}$ is absolutely convergent.

(11) Suppose that $a_n, b_n > 0$ for all n and that there exist a positive integer N and a real number $r > 0$ such that

$$n \geq N \quad \text{implies} \quad a_n b_n - a_{n+1} b_{n+1} \geq r a_n > 0.$$

   (a) Prove that the sequence $\{a_n b_n\}_{n=N}^{\infty}$ is strictly decreasing.

   (b) Prove that the whole sequence $\{a_n b_n\}$ is convergent.

   (c) Prove that $\sum_{n=N}^{\infty} (a_n b_n - a_{n+1} b_{n+1})$ is convergent with limit $a_N b_N - L$, where $L = \lim a_n b_n$.

   (d) Conclude that $\sum a_n$ is convergent. This is known as **Kummer's Test**.

(12) Let the setting be the same as in the previous exercise, and set

$$K_n = b_n - b_{n+1} \frac{a_{n+1}}{a_n}.$$

   (a) Suppose that $K = \liminf K_n > 0$ and that $\epsilon > 0$ with $K - \epsilon > 0$. Prove that there exists a positive integer N such that

$$n \geq N \quad \text{implies} \quad K_n \geq K - \epsilon.$$

   (b) If $K = \liminf K_n > 0$, prove that $\sum a_n$ is convergent.

   (c) Take $b_n = n - 1$ and set $R_n = K_n + 1$. Prove that

$$R_n = n \left(1 - \frac{a_{n+1}}{a_n}\right).$$

   (d) If $\liminf R_n > 1$, prove that $\sum a_n$ is convergent. This is known as **Raabe's Test**.

(13) Using Raabe's Test prove that the series

$$1 + x + \frac{x(x+1)}{2!} + \frac{x(x+1)(x+2)}{3!} + \frac{x(x+1)(x+2)(x+3)}{4!} + \cdots$$

is convergent if $x < 0$.

(14) Suppose that $a_n, b_n > 0$ and that $\sum_{n=1}^{\infty} a_n$ and $\sum_{n=1}^{\infty} b_n$ are both convergent. For each n define

$$c_n = \sum_{i=1}^{n} a_i b_{n+1-i} = a_1 b_n + a_2 b_{n-1} + \cdots + a_{n-1} b_2 + a_n b_1$$

Denote the nth partial sums of $\sum a_n$, $\sum b_n$, and $\sum c_n$ by A_n, B_n, and C_n, respectively.

(a) Prove that $C_n \le A_n B_n$ for all n.

(b) Prove that, for every integer $n \ge 1$, there exists an integer $N \ge n$ such that $A_n B_n \le C_N$.

(c) Prove that $\sum c_n$ is convergent with limit $(\sum a_n)(\sum b_n)$. The series $\sum c_n$ is called the **Cauchy product** of $\sum a_n$ and $\sum b_n$.

(15) Let $\{a_n\}$ and $\{b_n\}$ be two sequences, and let s_n be the nth partial sum of $\sum b_n$.

(a) For $n > m \ge 1$, prove **Abel's Formula**:

$$\sum_{k=m+1}^{n} a_k b_k = a_n s_n - a_{m+1} s_m + \sum_{k=m+1}^{n-1} (a_k - a_{k+1}) s_k.$$

(b) Suppose that $\{a_n\}$ is decreasing with $a_n \to 0$ and that $\{s_n\}$ is bounded. Prove that $\sum a_n b_n$ is convergent. This is known as **Dirichlet's Test**.

(16) Using Dirichlet's Test give another proof of the Alternating Series Test.

(17) Suppose that $\{a_n\}$ is a monotone convergent sequence and that $\sum b_n$ is convergent. Using Dirichlet's Test prove that $\sum a_n b_n$ is convergent. This is known as **Abel's Test**. Discuss why the monotone assumption on $\{a_n\}$ is necessary.

Chapter 4

Continuous Functions

In this chapter, we discuss continuous real valued functions defined on subsets of $\mathbb{R}$. Continuity is one of the most basic and important concepts about real valued functions. As we will discuss later in this book, wherever a function is differentiable, it is continuous as well. Also, continuous functions provide a large class of integrable functions. Continuity is closely related to limits of a function, which we discuss in sections 4.1 and 4.2.

Continuous functions are defined and discussed in section 4.3. In section 4.4 it is observed that when a continuous function f is defined on a closed bounded interval $[a, b]$, it attains both a maximum and a minimum on that interval. It also satisfies the intermediate value property. In section 4.5 we discuss a concept called uniform continuity that is stronger than continuity. A function that is continuous on a closed bounded interval is automatically uniformly continuous. In general, however, a continuous function is not necessarily uniformly continuous.

In section 4.6 we discuss monotone functions, the functional analogs of monotone sequences. It is shown that a monotone function on a bounded interval can have at most countably many points of discontinuity. If a strictly monotone function defined on a closed bounded interval is continuous, then so is its inverse function. In section 4.7 we discuss functions of bounded variation. Intuitively, a function of bounded variation is a function whose graph does not fluctuate too much. A function of bounded variation can be characterized as the difference of two increasing functions.

4.1   Limit Points

The limit of a function f at a point c is about the values of $f(x)$ when x is close, but not equal, to c. This leads naturally to the concept of limit points.

Definition 4.1. Let A be a non-empty subset of $\mathbb{R}$, and let x be a real number. Then we say that x is a **limit point** of A if for every $\epsilon > 0$, there exist infinitely many elements

$$a \in A \quad \text{such that} \quad 0 < |a - x| < \epsilon.$$

So x is *not* a limit point of A if and only if there exists an $\epsilon_0 > 0$ such that there are only finitely many elements $a \in A$ satisfying $0 < |a - x| < \epsilon_0$.

Note that a limit point of A is *not* required to be an element in A. Conversely, an element in A is not necessarily a limit point of A. Intuitively, if x is a limit point of A, then there are enough points in A that are close, but not equal, to x. In the following result, we provide two alternative characterizations of limit points.

Theorem 4.1. *Let A be a non-empty subset of $\mathbb{R}$, and let x be a real number. The following statements are equivalent.*

(1) *The number x is a limit point of A.*
(2) *For every $\epsilon > 0$, there exists an element*
$$a \in A \quad \text{such that} \quad 0 < |a - x| < \epsilon.$$
(3) *There exists a sequence*
$$\{a_n\} \subseteq A \smallsetminus \{x\} \quad \text{such that} \quad \lim a_n = x.$$

Proof. $(1) \Rightarrow (2)$ is immediate from the definition. To prove $(2) \Rightarrow (3)$, by the hypothesis of (2), there exists an element
$$a_1 \in A \quad \text{such that} \quad 0 < |a_1 - x| < 1.$$
Since $|a_1 - x| > 0$, there exists an element $a_2 \in A$ such that
$$0 < |a_2 - x| < \min\left\{|a_1 - x|, \frac{1}{2}\right\} \leq \frac{1}{2}.$$
Continuing this way we obtain a sequence $\{a_n\}$ of elements in A such that
$$0 < |a_n - x| < \min\left\{|a_{n-1} - x|, \frac{1}{n}\right\} \leq \frac{1}{n}$$
for $n \geq 2$. Since $\frac{1}{n} \to 0$, we conclude that $a_n \to x$.

To prove $(3) \Rightarrow (1)$, pick $\epsilon > 0$. Since we are now assuming that $\lim a_n = x$, there exists a positive integer N such that
$$n \geq N \quad \text{implies} \quad |a_n - x| < \epsilon.$$
There are infinitely many a_n with $n \geq N$, and $|a_n - x| > 0$ by the hypothesis of (3). So x is a limit point of A. $\qquad\square$

Corollary 4.1. *Let A be a non-empty subset of $\mathbb{R}$. Then a real number x is **not** a limit point of A if and only if there exists an $\epsilon_0 > 0$ such that*
$$a \in A \smallsetminus \{x\} \quad \text{implies} \quad |a - x| \geq \epsilon_0.$$

Proof. This is the negation of condition (2) in Theorem 4.1. $\qquad\square$

Example 4.1. The set $A = \{\frac{1}{n} : n \in \mathbb{Z}_+\}$ has $0 \notin A$ as its only limit point. In fact, since $\frac{1}{n} \to 0$, it follows that 0 is a limit point of A. On other other hand, no real number $x \neq 0$ is a limit point of A by Corollary 4.1.

Example 4.2. The set $\mathbb{Q}$ of rational numbers has $\mathbb{R}$ as its set of limit points. Indeed, if x is an arbitrary real number and $\epsilon > 0$, then there exists a rational number a such that $x < a < x + \epsilon$ (Theorem 1.3). This is equivalent to $0 < a - x < \epsilon$, which, by condition (2) in Theorem 4.1, implies that x is a limit point of $\mathbb{Q}$.

4.1.1 *Exercises*

(1) Prove that a non-empty finite set has no limit points.
(2) Prove that every point c in an interval I is a limit point of I.
(3) Suppose that $a < b$. Prove that the set of limit points of the open interval (a, b) is the closed interval $[a, b]$.
(4) Prove that a closed interval is equal to its set of limit points.
(5) Prove that c is a limit point of a non-empty subset A in $\mathbb{R}$ if and only if at least one of the following conditions is satisfied:

    (a) c is a limit point of $\{a \in A : a < c\}$.
    (b) c is a limit point of $\{a \in A : a > c\}$.

4.2 Limits of Functions

The purpose of this section is to discuss limits of a function. From now on, unless otherwise specified, whenever we say *function*, we mean a function whose domain $Dom(f)$ is a subset of $\mathbb{R}$ and whose target is $\mathbb{R}$.

4.2.1 *Limits*

Definition 4.2. Let $f : A \to \mathbb{R}$ be a function, and let c be a limit point of A. Suppose that L is a real number. We say that the **limit of** f **at** c **is** L if for every sequence $\{a_n\}$ in $A \setminus \{c\}$,

$$\lim a_n = c \quad \text{implies} \quad \lim f(a_n) = L.$$

In this case, we write

$$\lim_{x \to c} f = L \quad \text{or} \quad f \to L \text{ as } x \to c,$$

and say that f **converges to** L as x approaches c. If no such L exists, then we say that the limit of f at c does not exist.

In particular, the limit of f at c is *not* L if there exists a sequence $\{a_n\}$ in $A \setminus \{c\}$ with $\lim a_n = c$ such that $\{f(a_n)\}$ does not converge to L. From now on, whenever we use the symbol $\lim_{x \to c} f$, the point c is assumed to be a limit point of the domain of the function f.

Observe that the limit of f at a point is really about the convergence of certain sequences $\{f(a_n)\}$, which we discussed in the previous two chapters. Many results about sequences can be translated into the context of limits of functions, some of which are in the exercises. The reader should be careful that in order to have $\lim_{x \to c} f = L$, the convergence $f(a_n) \to L$ must hold for *every* sequence in $A \setminus \{c\}$ with $a_n \to c$. It is not enough to check $f(a_n) \to L$ for just one sequence $\{a_n\}$.

Be careful that $\lim_{x \to c} f$, whether it exists or not, is *not* about the value of $f(c)$. In particular, $\lim_{x \to c} f$ may exist even though f is not defined at c. Even if f is defined at c, the limit $\lim_{x \to c} f$ is not necessarily equal to $f(c)$.

Example 4.3. Consider the function $f(x) = \frac{6x^2 - x - 1}{2x - 1}$ with domain $\mathbb{R} \smallsetminus \{\frac{1}{2}\}$. We want to compute $\lim_{x \to \frac{1}{2}} f$, if it exists. Pick any sequence $\{a_n\}$ in $\mathbb{R} \smallsetminus \{\frac{1}{2}\}$ that converges to $\frac{1}{2}$. Since $a_n \neq \frac{1}{2}$ for any n, we have

$$f(a_n) = \frac{6a_n^2 - a_n - 1}{2a_n - 1} = 3a_n + 1 \to 3\left(\frac{1}{2}\right) + 1 = \frac{5}{2}.$$

This shows that $\lim_{x \to \frac{1}{2}} f = \frac{5}{2}$, even though f is not defined at $\frac{1}{2}$.

Example 4.4. Consider the function $f: \mathbb{R} \to \mathbb{R}$ defined as

$$f(x) = \begin{cases} \sin\left(\frac{1}{x}\right) & \text{if } x \neq 0, \\ 0 & \text{if } x = 0. \end{cases}$$

The reader should sketch the graph of f to visualize its behavior when x is close to 0. On any open interval containing 0, the graph of f fluctuates between -1 and 1 infinitely often. From its graph, one can guess that $\lim_{x \to 0} f$ does not exist. To prove it, consider the sequence with $a_n = \frac{1}{(n - \frac{1}{2})\pi}$. We have $a_n \to 0$, and

$$f(a_n) = \sin\left(n - \frac{1}{2}\right)\pi = \begin{cases} 1 & \text{if } n \text{ is odd}, \\ -1 & \text{if } n \text{ is even}. \end{cases}$$

So the sequence $\{f(a_n)\}$ is divergent, showing that $\lim_{x \to 0} f$ does not exist.

4.2.2   Uniqueness of Limits

As one can expect from the uniqueness of limits of sequences, the limit of a function at a point, if it exists, is unique.

Theorem 4.2. *Let $f: A \to \mathbb{R}$ be a function, and let c be a limit point of A. If the limit of f at c exists, then it is unique.*

Proof.   Suppose that $\lim_{x \to c} f = L$ and $\lim_{x \to c} f = L'$. We want to show that $L = L'$. Pick any sequence $\{a_n\}$ in $A \smallsetminus \{c\}$ with $\lim a_n = c$. Such a sequence exists by Theorem 4.1. Then

$$\lim f(a_n) = L \quad \text{and} \quad \lim f(a_n) = L'.$$

Since the limit of a convergent sequence is unique (Theorem 2.1), we conclude that $L = L'$.   □

4.2.3 ϵ-δ Characterization

Next we provide an alternative characterization of limits of a function in a form that is closer to the definition of uniform continuity, which will be discussed in the next section.

Theorem 4.3. *Let $f: A \to \mathbb{R}$ be a function, c be a limit point of A, and L be a real number. Then*

$$\lim_{x \to c} f = L$$

if and only if for every $\epsilon > 0$, there exists $\delta > 0$ such that

$$0 < |x - c| < \delta \text{ with } x \in A \quad \text{implies} \quad |f(x) - L| < \epsilon.$$

Proof. For the "only if" part, suppose that $\lim_{x \to c} f = L$, and let $\epsilon > 0$ be given. We prove the existence of the required $\delta > 0$ by contradiction, so assume that no $\delta > 0$ satisfies the stated condition. For $\delta_1 = 1$, there must be an $a_1 \in A \setminus \{c\}$ with

$$|a_1 - c| < 1 \quad \text{and} \quad |f(a_1) - L| \geq \epsilon.$$

Next, for $\delta_2 = \min\{\frac{1}{2}, |a_1 - c|\} > 0$, there must be an $a_2 \in A \setminus \{c\}$ with

$$|a_2 - c| < \delta_2 \quad \text{and} \quad |f(a_2) - L| \geq \epsilon.$$

Continuing this way we obtain a sequence $\{a_n\}$ in $A \setminus \{c\}$ with

$$|a_n - c| < \frac{1}{n} \quad \text{and} \quad |f(a_n) - L| \geq \epsilon$$

for each n. Since $a_n \to c$ and $\{f(a_n)\}$ does not converge to L, we conclude that L is not the limit of f at c, which is a contradiction. This proves the "only if" part.

For the "if" part, suppose that the stated ϵ-δ condition is satisfied. Suppose that $\{a_n\}$ is an arbitrary sequence in $A \setminus \{c\}$ with $\lim a_n = c$. To show that $f(a_n) \to L$, let $\epsilon > 0$ be given. Using the $\delta > 0$ from the ϵ-δ condition, we know that there exists a positive integer N such that

$$n \geq N \quad \text{implies} \quad 0 < |a_n - c| < \delta.$$

The ϵ-δ condition now implies that

$$|f(a_n) - L| < \epsilon$$

for $n \geq N$. This shows that $\lim_{x \to c} f = L$. $\qquad \square$

The ϵ-δ condition can be understood as follows. Given an $\epsilon > 0$, the value of $f(x)$ can be made ϵ-close to L, provided that $x \in A \setminus \{c\}$ is chosen to be δ-close to c. Be careful that *first* $\epsilon > 0$ is given, and *then* $\delta > 0$ is chosen to make certain inequality true. The value of δ, in general, depends on both c and ϵ.

Example 4.5. Consider the function $f: \mathbb{R} \to \mathbb{R}$ defined as

$$f(x) = \begin{cases} x^2 \sin\left(\frac{1}{x}\right) & \text{if } x \neq 0, \\ 2 & \text{if } x = 0. \end{cases}$$

From the graph of f, one can see that $\lim_{x \to 0} f$ seems to be 0. We will prove this using the ϵ-δ characterization of limits. For $x \neq 0$ we have

$$|f(x) - 0| = \left| x^2 \sin\left(\frac{1}{x}\right) \right| \leq x^2,$$

since $|\sin(x)| \leq 1$ for all x. Given $\epsilon > 0$ we take $\delta = \sqrt{\epsilon} > 0$. Then

$$0 < |x - 0| < \delta \quad \text{implies} \quad |f(x) - 0| \leq x^2 < \delta^2 = \epsilon.$$

This shows that $\lim_{x \to 0} f = 0$, *which is different from* $f(0)$.

Recall that a convergent sequence is bounded (Theorem 2.2). If the limit of a function f exists at a point c, then it makes sense that f should be bounded near c. This is made precise in the following result.

Theorem 4.4. *Let $f: A \to \mathbb{R}$ be a function, and let c be a limit point of A. Suppose that $\lim_{x \to c} f$ exists. Then there exist real numbers $\delta > 0$ and $M > 0$ such that*

$$|x - c| < \delta \text{ with } x \in A \quad \text{implies} \quad |f(x)| \le M.$$

Proof. Say $\lim_{x \to c} f$ is L. By Theorem 4.3, given $\epsilon = 1$ there exists $\delta > 0$ such that

$$0 < |x - c| < \delta \text{ with } x \in A \quad \text{implies} \quad |f(x) - L| < 1.$$

This in turn implies

$$|f(x)| < |L| + 1.$$

We now take

$$M = \begin{cases} \max\{|f(c)|, |L| + 1\} & \text{if } c \in A, \\ |L| + 1 & \text{if } c \notin A. \end{cases}$$

Then we have $|f(x)| \le M$ whenever $|x - c| < \delta$ with $x \in A$. $\square$

4.2.4 *One-Sided Limits*

For some purposes, such as the discussion of monotone functions in section 4.6, it is convenient to consider limits when x approaches a point c from below or above. First we define limit from below.

Definition 4.3 (Left-Hand Limits). *Let $f: A \to \mathbb{R}$ be a function, c be a limit point of*

$$A_{<c} = \{x \in A : x < c\},$$

and L be a real number.

(1) *We say that the **left-hand limit of f at c is L** if for every sequence $\{a_n\}$ in $A_{<c}$,*

$$\lim a_n = c \quad \text{implies} \quad \lim f(a_n) = L.$$

In this case, we write $\lim_{x \to c^-} f = L$, and say that $\lim_{x \to c^-} f$ exists.

(2) *We say that the **left-hand limit of f at c is ∞** (or $-\infty$) if for every sequence $\{a_n\}$ in $A_{<c}$,*

$$\lim a_n = c \quad \text{implies} \quad \lim f(a_n) = \infty \quad (or \; -\infty).$$

In this case, we write $\lim_{x \to c^-} f = \infty$ (or $-\infty$).

The definition of limit from above is similar.

Definition 4.4 (Right-Hand Limits). *Let $f: A \to \mathbb{R}$ be a function, c be a limit point of*

$$A_{>c} = \{x \in A : x > c\},$$

and L be a real number.

*(1) We say that the **right-hand limit of f at c is L** if for every sequence $\{a_n\}$ in $A_{>c}$,*

$$\lim a_n = c \quad implies \quad \lim f(a_n) = L.$$

In this case, we write $\lim_{x \to c^+} f = L$, and say that $\lim_{x \to c^+} f$ exists.
*(2) We say that the **right-hand limit of f at c is ∞** (or $-\infty$) if for every sequence $\{a_n\}$ in $A_{>c}$,*

$$\lim a_n = c \quad implies \quad \lim f(a_n) = \infty \quad (or\ -\infty).$$

In this case, we write $\lim_{x \to c^+} f = \infty$ (or $-\infty$).

Left-hand limits and right-hand limits are both called one-sided limits. If the one-sided limits $\lim_{x \to c^-} f$ and $\lim_{x \to c^+} f$ are considered, then it is assumed that c is a limit point of $\{x \in Dom(f) : x < c\}$ and $\{x \in Dom(f) : x > c\}$, respectively.

As one can expect, $\lim_{x \to c} f = L$ if and only if the two one-sided limits are equal to L. Moreover, one-sided limits can be characterized in terms of variations of the ϵ-δ condition in Theorem 4.3. These statements are left as exercises for the reader.

4.2.5 Exercises

(1) Write down an ϵ-δ characterization of $\lim_{x \to c} f \neq L$.
(2) Suppose that $f, g: A \to \mathbb{R}$ are functions, $\lim_{x \to c} f = L$, $\lim_{x \to c} g = M$, and $a \in \mathbb{R}$.

    (a) Prove that $\lim_{x \to c} af = aL$, where $(af)(x) = af(x)$ for $x \in A$.
    (b) Prove that $\lim_{x \to c}(f+g) = L+M$, where $(f+g)(x) = f(x)+g(x)$ for $x \in A$.
    (c) Prove that $\lim_{x \to c}(fg) = LM$, where $(fg)(x) = f(x)g(x)$ for $x \in A$.
    (d) Suppose, in addition, that $g(x) \neq 0$ for any $x \in A$ and $M \neq 0$. Prove that $\lim_{x \to c} \frac{f}{g} = \frac{L}{M}$, where $(\frac{f}{g})(x) = \frac{f(x)}{g(x)}$ for $x \in A$.

(3) Let $p(x)$ be a polynomial. Prove that $\lim_{x \to c} p = p(c)$.
(4) Given a function $f: A \to \mathbb{R}$, define $|f|: A \to \mathbb{R}$ by $|f|(x) = |f(x)|$ for $x \in A$.

    (a) If $\lim_{x \to c} f = L$, prove that $\lim_{x \to c} |f| = |L|$.
    (b) Give an example in which $\lim_{x \to c} |f|$ exists, but $\lim_{x \to c} f$ does not exist.

(5) Given a function $f: A \to \mathbb{R}$ satisfying $f(x) \geq 0$ for all $x \in A$, define $\sqrt{f}: A \to \mathbb{R}$ by $\sqrt{f}(x) = \sqrt{f(x)}$ for $x \in A$.

    (a) If $\lim_{x \to c} f = L$, prove that $\lim_{x \to c} \sqrt{f} = \sqrt{L}$.
    (b) Is the converse of the previous part true?

(6) Suppose that $a \leq f(x) \leq b$ for all $x \in Dom(f)$ for some real numbers a and b. If $\lim_{x \to c} f = L$, prove that $a \leq L \leq b$.

(7) Suppose that $f, g, h \colon A \to \mathbb{R}$ are functions such that $f(x) \le g(x) \le h(x)$ for all $x \in A$. If $\lim_{x \to c} f = L = \lim_{x \to c} h$, prove that $\lim_{x \to c} g = L$.

(8) Use the ϵ-δ characterization of limits to prove the following statements.

    (a) $\lim_{x \to 2} \sqrt{x} = \sqrt{2}$.

    (b) $\lim_{x \to 1} (x^2 + 1)^{-1} = 1/2$.

    (c) $\lim_{x \to 1} \frac{2x}{3x^2 + 1} = 1/2$.

(9) Determine if $\lim_{x \to 0} x \sin(\frac{1}{x^2})$ exists or not.

(10) Suppose that $\lim_{x \to c} f > 0$. Prove that there exists $\delta > 0$ such that $|x - c| < \delta$ with $x \in Dom(f) \smallsetminus \{c\}$ implies $f(x) > 0$.

(11) Prove that $\lim_{x \to c} f = L$ if and only if both $\lim_{x \to c^-} f = L$ and $\lim_{x \to c^+} f = L$

(12) (a) Prove that $\lim_{x \to c^-} f = L$ if and only if for every $\epsilon > 0$, there exists $\delta > 0$ such that $0 < c - x < \delta$ implies $|f(x) - L| < \epsilon$.

    (b) Prove that $\lim_{x \to c^+} f = L$ if and only if for every $\epsilon > 0$, there exists $\delta > 0$ such that $0 < x - c < \delta$ implies $|f(x) - L| < \epsilon$.

(13) (a) Prove that $\lim_{x \to c^-} f = \infty$ if and only if for every real number $M > 0$, there exists $\delta > 0$ such that $0 < c - x < \delta$ implies $f(x) > M$.

    (b) Prove that $\lim_{x \to c^-} f = -\infty$ if and only if for every real number $M < 0$, there exists $\delta > 0$ such that $0 < c - x < \delta$ implies $f(x) < M$.

    (c) Formulate and prove similar statements for $\lim_{x \to c^+} f = \pm\infty$.

(14) In each case, give an example that has the stated properties.

    (a) Both $\lim_{x \to c^-} f$ and $\lim_{x \to c^+} f$ exist, but they are not equal.

    (b) $\lim_{x \to c^-} f$ exists, but $\lim_{x \to c^+} f$ does not exist.

    (c) $\lim_{x \to c^-} f = L \in \mathbb{R}$ and $\lim_{x \to c^+} f = \infty$.

    (d) $\lim_{x \to c^-} f = \infty$ and $\lim_{x \to c^+} f = -\infty$.

4.3 Continuity

In this section we study continuous functions. When a continuous function is defined on a closed bounded interval, we will show in section 4.4 that it attains both a maximum and a minimum on that interval. Moreover, it satisfies the Intermediate Value Theorem. In section 4.5, we will discuss uniform continuity, a condition that is stronger than continuity. We will see that a continuous function on a closed bounded interval is uniformly continuous.

4.3.1 *Sequential Definition of Continuity*

For a function f to be continuous at a point $a \in Dom(f)$, what it should mean is that $f(x)$ can be made as close to $f(a)$ as one wants, provided that x is sufficiently close to a. As in the case of limits of a function, there are two equivalent ways to express this concept, one in terms of sequences and the other in terms of ϵ-δ. We begin with the first one.

Definition 4.5. Let $f: A \to \mathbb{R}$ be a function, and let $a \in A$. We say that f **is continuous at** a if for every sequence $\{a_n\}$ in A,

$$\lim a_n = a \quad \text{implies} \quad \lim f(a_n) = f(a).$$

If f is not continuous at a, we say that f is **discontinuous at** a. If f is continuous at each point in a non-empty subset $B \subseteq A$, we say that f is continuous on B. We call $f: A \to \mathbb{R}$ a **continuous function** if f is continuous on A.

In particular, a function f is *not* continuous at a if there exists a sequence $\{a_n\}$ in A with $\lim a_n = a$ such that $\{f(a_n)\}$ does not converge to $f(a)$.

Notice that whether f is continuous at a has everything to do with the value of $f(a)$, unlike the case of $\lim_{x \to a} f$. In the above definition, it is not required that a be a limit point of A. If a is not a limit point of A, then f is automatically continuous at a (Exercise (7) on page 111).

On the other hand, if a is a limit point of A, then a comparison with Definition 4.2 shows that f is continuous at a if and only if

$$\lim_{x \to a} f = f(a). \tag{4.1}$$

The reader should write down a proof of this assertion.

Example 4.6. A polynomial $p(x)$ is continuous on all of $\mathbb{R}$. Indeed, the domain of p is $\mathbb{R}$, so every point a is a limit point of $Dom(p)$. Moreover, we have $\lim_{x \to a} p = p(a)$ for every a (Exercise (3) on page 89).

One should *not* expect that functions in general have simple continuity behaviors. The next two examples illustrate that continuity can be a delicate issue.

Example 4.7. Consider the function $\chi_{\mathbb{Q}}: \mathbb{R} \to \mathbb{R}$ defined as

$$\chi_{\mathbb{Q}}(x) = \begin{cases} 1 & \text{if } x \in \mathbb{Q}, \\ 0 & \text{if } x \notin \mathbb{Q}. \end{cases} \tag{4.2}$$

This is called the **characteristic function** of $\mathbb{Q}$. We claim that $\chi_{\mathbb{Q}}$ is discontinuous at every point $a \in \mathbb{R}$. Indeed, first suppose that $a \in \mathbb{Q}$, so $\chi_{\mathbb{Q}}(a) = 1$. Pick any sequence of irrational numbers $\{a_n\}$ converging to a, such as $a_n = a + \frac{\sqrt{2}}{n}$. Then

$$\chi_{\mathbb{Q}}(a_n) = 0 \to 0 \neq \chi_{\mathbb{Q}}(a),$$

showing that $\chi_{\mathbb{Q}}$ is discontinuous at any $a \in \mathbb{Q}$. On the other hand, suppose that $b \notin \mathbb{Q}$, so $\chi_{\mathbb{Q}}(b) = 0$. Pick any sequence of rational numbers $\{b_n\}$ converging to b. The existence of such a sequence is guaranteed by Theorem 1.3. Then

$$\chi_{\mathbb{Q}}(b_n) = 1 \to 1 \neq \chi_{\mathbb{Q}}(b),$$

so $\chi_{\mathbb{Q}}$ is discontinuous at any $b \notin \mathbb{Q}$.

Example 4.8. Consider the function $f: \mathbb{R} \to \mathbb{R}$ defined as

$$f(x) = \begin{cases} x & \text{if } x \in \mathbb{Q}, \\ 0 & \text{if } x \notin \mathbb{Q}. \end{cases}$$

Note that this function is the product $f = x \cdot \chi_{\mathbb{Q}}$. We claim that f is continuous at 0 and is discontinuous at any $x \neq 0$. Indeed, since

$$|f(x) - f(0)| = |f(x)| \leq |x|,$$

if $a_n \to 0$, then $f(a_n) \to 0 = f(0)$. This shows that f is continuous at 0. On the other hand, if $x \neq 0$ and $x \in \mathbb{Q}$, then we pick a sequence $\{a_n\}$ of irrational numbers with $a_n \to x$. Then

$$f(a_n) = 0 \to 0 \neq x = f(x).$$

If $x \notin \mathbb{Q}$, then we pick a sequence $\{b_n\}$ of rational numbers with $b_n \to x$. Then

$$f(b_n) = b_n \to x \neq 0 = f(x).$$

Thus, f is discontinuous at any $x \neq 0$.

4.3.2 ϵ-δ *Characterization of Continuity*

Since limits can be characterized with an ϵ-δ condition, the same can be expected of continuity.

Theorem 4.5. *Let $f: A \to \mathbb{R}$ be a function, and let $a \in A$. Then f is continuous at a if and only if for ever $\epsilon > 0$, there exists $\delta > 0$ such that*

$$|x - a| < \delta \text{ with } x \in A \quad \text{implies} \quad |f(x) - f(a)| < \epsilon.$$

The proof of this theorem is almost identical to that of Theorem 4.3 and will be left as an exercise for the reader.

Example 4.9. Consider the function $f(x) = x^{-1}$ defined on $(0, 1)$. We show that it is continuous on $(0, 1)$ using the ϵ-δ characterization of continuity. Pick any point $a \in (0, 1)$, and let $\epsilon > 0$ be given. We need to estimate

$$|f(x) - f(a)| = \left| \frac{1}{x} - \frac{1}{a} \right| = \frac{|a - x|}{xa}.$$

To make this less than ϵ, we need a suitable lower bound of x. If we take

$$\delta = \min \left\{ \frac{a^2 \epsilon}{2}, \frac{a}{2} \right\},$$

then

$$|x - a| < \delta \text{ with } x \in (0, 1) \quad \text{implies} \quad x > \frac{a}{2}.$$

For such x, we have

$$|f(x) - f(a)| = \frac{|a - x|}{xa} < \frac{2|a - x|}{a^2} < \frac{2\delta}{a^2} \leq \epsilon.$$

By Theorem 4.5 this shows that f is continuous at a. Note that this δ is dependent on both a and ϵ. A different δ is needed if either the point a or ϵ is changed.

4.3.3 *Exercises*

(1) Prove Theorem 4.5.

(2) Suppose a is a limit point of the domain of f. Prove that f is continuous at a if and only if (4.1) holds.

(3) Suppose that $f, g: A \to \mathbb{R}$ are continuous at $a \in A$.

    (a) Prove that cf is continuous at a, where c is any real number and $(cf)(x) = cf(x)$ for $x \in A$.

    (b) Prove that $f + g$ is continuous at a, where $(f + g)(x) = f(x) + g(x)$ for $x \in A$.

    (c) Prove that fg is continuous at a, where $(fg)(x) = f(x)g(x)$ for $x \in A$.

    (d) Suppose, in addition, that $g(a) \neq 0$. Prove that $\frac{f}{g}$ is continuous at a.

(4) Suppose that $f: A \to \mathbb{R}$ is continuous at a. If $|f|: A \to \mathbb{R}$ is defined as $|f|(x) = |f(x)|$ for $x \in A$, prove that $|f|$ is continuous at a. Is the converse true?

(5) Suppose that $f: A \to \mathbb{R}$ is continuous at a and $f(x) \geq 0$ for all $x \in A$. If $\sqrt{f}: A \to \mathbb{R}$ is defined as $\sqrt{f}(x) = \sqrt{f(x)}$ for $x \in A$, prove that $\sqrt{f}$ is continuous at a. Is the converse true?

(6) Suppose that f and g are functions with $Ran(f) \subseteq Dom(g)$. Define the composition of f and g by $(g \circ f)(x) = g(f(x))$ for $x \in Dom(f)$. If f is continuous at a and g is continuous at $f(a)$, prove that $g \circ f$ is continuous at a.

(7) Suppose that f is continuous at a

    (a) If $f(a) > 0$, prove that there exists $\delta > 0$ such that $|x - a| < \delta$ with $x \in Dom(f)$ implies $f(x) > 0$.

    (b) If $f(a) < 0$, prove that there exists $\delta > 0$ such that $|x - a| < \delta$ with $x \in Dom(f)$ implies $f(x) < 0$.

(8) For a subset $S \subseteq \mathbb{R}$, define its **characteristic function** $\chi_S: \mathbb{R} \to \mathbb{R}$ by

$$\chi_S(x) = \begin{cases} 1 & \text{if } x \in S, \\ 0 & \text{if } x \notin S. \end{cases}$$

Determine the points of continuity of $\chi_{\mathbb{Z}}$, $\chi_{\mathbb{I}}$, and $x \cdot \chi_{\mathbb{I}}$.

4.4 Extreme and Intermediate Value Theorems

One reason why continuous functions are important is that they have certain good properties when defined on an interval. The first such result says that a continuous function defined on a closed bounded interval is bounded and attains both a maximum and a minimum. To be precise, we need the following definition.

Definition 4.6. A function $f: A \to \mathbb{R}$ is said to be **bounded** if there exists

$$M > 0 \quad \text{such that} \quad |f(x)| \leq M$$

for all $x \in A$.

Theorem 4.6 (Extreme Value Theorem). *Let* $f:[a,b] \to \mathbb{R}$ *be a continuous function. Then* f *is bounded. Moreover, there exist* α *and* β *in* $[a,b]$ *such that*

$$f(\alpha) \le f(x) \le f(\beta)$$

for all $x \in [a,b]$.

Ideas of Proof. That f is bounded will be shown to be a consequence of the Bolzano-Weierstrass Theorem and the fact that a convergent sequence is bounded. The existence of α is shown by constructing a sequence in $[a,b]$ whose image under f converges to the infimum of f on $[a,b]$.

Proof.   We first prove that f is bounded by contradiction. If f is not bounded, then there exists a sequence $\{a_n\}$ in $I = [a,b]$ such that $|f(a_n)| \to \infty$. Since $\{a_n\}$ is bounded, it has a convergent subsequence $\{a_{n_k}\}$ by the Bolzano-Weierstrass Theorem 2.8. We also have $|f(a_{n_k})| \to \infty$. Since f is continuous and $\{a_{n_k}\}$ is convergent, the sequence $\{f(a_{n_k})\}$ is convergent, and hence bounded (Theorem 2.2). This is a contradiction, so f is bounded.

Next we show the existence of $\alpha \in I$ such that $f(\alpha) \le f(x)$ for all $x \in I$. We just proved in the previous paragraph that the set $\{f(x) : x \in I\}$ is bounded, so $m = \inf\{f(x) : x \in I\}$ is a real number. It suffices to show that $f(\alpha) = m$ for some $\alpha \in I$. For each $n \in \mathbb{Z}_+$, there must exist $x_n \in I$ such that

$$m \le f(x_n) < m + \frac{1}{n},$$

so we have $f(x_n) \to m$. By the Bolzano-Weierstrass Theorem 2.8 again, $\{x_n\}$ has a convergent subsequence $\{x_{n_k}\}$ with limit $\alpha \in I$ because I is a closed bounded interval. Since f is continuous at α, it follows that $f(x_{n_k}) \to f(\alpha)$. But $\{f(x_{n_k})\}$ is a subsequence of $\{f(x_n)\}$, so $f(x_{n_k}) \to m$ (Theorem 2.6). The uniqueness of the limit of a convergent sequence (Theorem 2.1) now implies that $m = f(\alpha)$.

The existence of β is proved similarly by considering $M = \sup\{f(x) : x \in I\}$ instead of m. The details are left to the reader as an exercise. $\square$

The reader should be careful that the hypothesis of I being a closed bounded interval cannot be omitted from the Extreme Value Theorem.

The next result says that a continuous function f defined on a closed bounded interval $[a,b]$ has the intermediate value property: If r is strictly between $f(a)$ and $f(b)$, then $r = f(x)$ for some $x \in [a,b]$. This seems pretty obvious. However, it does take a bit of work to give a rigorous proof of this fact.

Theorem 4.7 (Intermediate Value Theorem). *Let* $f:[a,b] \to \mathbb{R}$ *be a continuous function such that* $f(a) \ne f(b)$. *If* r *lies strictly between* $f(a)$ *and* $f(b)$, *then there exists a point*

$$x_0 \in [a,b] \quad \text{such that} \quad r = f(x_0).$$

Ideas of Proof. We show the existence of x_0 by considering the supremum of the set of points in $[a, b]$ whose images under f is $< r$.

Proof. There are two possibilities. We have either

$$f(a) < r < f(b) \quad \text{or} \quad f(b) < r < f(a).$$

We will consider the first case and leave the similar second case to the reader as an exercise. Consider the set

$$S = \{x \in [a, b] : f(x) < r\},$$

which is non-empty because $a \in S$. Its least upper bound $x_0 = \sup S$ is still in $[a, b]$. We will prove that $f(x_0) = r$.

For each $n \in \mathbb{Z}_+$, there exists $x_n \in S$ such that

$$x_0 - \frac{1}{n} < x_n \le x_0,$$

so we have $x_n \to x_0$. By the continuity of f, we have $f(x_n) \to f(x_0)$, so

$$f(x_0) \le r$$

because each $f(x_n) < r$. It remains to show that $f(x_0) \ge r$ as well. Consider

$$w_n = \min\left\{b, x_0 + \frac{1}{n}\right\} \in [x_0, b] \subseteq [a, b].$$

Each w_n satisfies

$$x_0 \le w_n \le x_0 + \frac{1}{n},$$

so $w_n \to x_0$. The continuity of f implies that $f(w_n) \to f(x_0)$. Now since

$$x_0 + \frac{1}{n} > x_0 = \sup S \quad \text{and} \quad f(b) > r,$$

we know that $w_n \notin S$. Therefore, we have $f(w_n) \ge r$ for each n. This implies that

$$f(x_0) \ge r,$$

as desired. $\square$

It should be emphasized that the previous two theorems give *sufficient* conditions, namely, being continuous on a closed bounded interval, that guarantee that a function has the extreme and intermediate value properties. However, being continuous on a closed bounded interval is not necessary for a function to have such properties. For example, as we will see in the next chapter, the derivative of a function need not be continuous. However, it does have the intermediate value property.

4.4.1   *Exercises*

(1) Finish the proof of Theorem 4.6 by proving the existence of β.
(2) Finish the proof of Theorem 4.7 by proving the case $f(b) < r < f(a)$.
(3) Let $f: I \to \mathbb{R}$ be a continuous function, where I is a closed bounded interval. Prove that $f(I) = \{f(x) : x \in I\}$ is either a single point or a closed bounded interval.
(4) Let $f: I \to \mathbb{R}$ be a continuous function, where I is an interval. Prove that $f(I)$ is either a single point or an interval.
(5) Let $f: [a, b] \to \mathbb{R}$ be a continuous function such that $f(a)f(b) < 0$. Prove that $f(c) = 0$ for some $c \in (a, b)$.
(6) Let $f: [0, 1] \to [0, 1]$ be a continuous function. Prove that it has a **fixed point**, i.e., a point $a \in [0, 1]$ such that $f(a) = a$.
(7) Show by a counterexample that the conclusions of the Extreme Value Theorem do not have to hold if the domain of the continuous function is an open bounded interval.
(8) Give an example of a function $f: [a, b] \to \mathbb{R}$ that is not continuous somewhere in $[a, b]$ but that satisfies the conclusions of both the Extreme Value Theorem and the Intermediate Value Theorem.

4.5   Uniform Continuity

In the ϵ-δ characterization of continuity, it makes sense that smaller values of $\delta > 0$ are needed if $\epsilon > 0$ is getting smaller. For a fixed $\epsilon > 0$, is it possible that one δ will work for all the points a in $Dom(f)$? We need the following concept to make this idea precise.

Definition 4.7. A function $f: A \to \mathbb{R}$ is said to be **uniformly continuous** on A if for every $\epsilon > 0$, there exists $\delta > 0$ such that

$$|a_0 - a_1| < \delta \text{ with } a_0, a_1 \in A \quad \text{implies} \quad |f(a_0) - f(a_1)| < \epsilon.$$

So a function $f: A \to \mathbb{R}$ is *not* uniformly continuous on A if and only if there exists $\epsilon_0 > 0$ such that for every $\delta > 0$, there exist

$$a_0, a_1 \in A \text{ with } |a_0 - a_1| < \delta \quad \text{such that} \quad |f(a_0) - f(a_1)| \geq \epsilon_0.$$

To explain the definition a bit further, a function $f: A \to \mathbb{R}$ is uniformly continuous if and only if the following is true. For each $\epsilon > 0$, there exists $\delta > 0$ such that, given *any* two points a_0 and a_1 in A that are within δ of each other, their images under f are guaranteed to be within ϵ of each other. It is in this sense that this δ works for all the points in A.

We now show that uniform continuity implies continuity.

Theorem 4.8. *Let $f: A \to \mathbb{R}$ be uniformly continuous on A. Then f is continuous on A.*

Proof. Pick any point $a \in A$. To prove that f is continuous at a, let $\epsilon > 0$ be given. Since f is uniformly continuous on A, there exists $\delta > 0$ such that

$$|a_0 - a_1| < \delta \text{ with } a_0, a_1 \in A \quad \text{implies} \quad |f(a_0) - f(a_1)| < \epsilon.$$

In particular, we have that

$$|x - a| < \delta \text{ with } x \in A \quad \text{implies} \quad |f(x) - f(a)| < \epsilon,$$

so f is continuous at a. $\qquad\qquad\square$

The converse of the above theorem is false. In other words, there are continuous functions that are not uniformly continuous. The next example gives one such function.

Example 4.10. Consider once again the function $f(x) = x^{-1}$ with domain $(0,1)$. We showed in Example 4.9 that f is continuous on $(0,1)$. Now we show that f is *not* uniformly continuous on $(0,1)$. With $\epsilon_0 = 1$ we will show that no $\delta > 0$ can satisfy the condition in Definition 4.7. Indeed, given any $\delta > 0$, we can take

$$a_0 = \min\left\{\delta, \frac{1}{2}\right\} \quad \text{and} \quad a_1 = \frac{a_0}{2}.$$

Then

$$|a_0 - a_1| = \frac{a_0}{2} \le \frac{\delta}{2} < \delta.$$

Moreover, we have

$$\left|\frac{1}{a_0} - \frac{1}{a_1}\right| = \frac{1}{a_0} > 1 = \epsilon_0.$$

So f is not uniformly continuous on $(0,1)$.

Theorem 4.8 and Example 4.10 together imply that uniform continuity is strictly stronger than continuity in general. This leads naturally to the following question. Is there a simple additional condition that can guarantee that a continuous function is uniformly continuous? The following result provides one simple answer to this question.

Theorem 4.9. *Let $f: I \to \mathbb{R}$ be a continuous function, where $I = [a,b]$ is a closed bounded interval. Then f is uniformly continuous on I.*

Ideas of Proof. We will show that, if f is not uniformly continuous, then there are two sequences whose images under f are separated by a certain fixed distance. But they also have subsequences converging to the same point. The continuity of f will then lead to a contradiction.

Proof. We prove this by contradiction. If f is not uniformly continuous on I, then there exists $\epsilon_0 > 0$ such that for every $\delta > 0$, there exist $a_0, a_1 \in I$ with

$$|a_0 - a_1| < \delta \quad \text{and} \quad |f(a_0) - f(a_1)| \ge \epsilon_0.$$

So for each $n \in \mathbb{Z}_+$, there exist $x_n, w_n \in I$ such that

$$|x_n - w_n| < \frac{1}{n} \quad \text{and} \quad |f(x_n) - f(w_n)| \geq \epsilon_0. \tag{4.3}$$

Since the sequence $\{x_n\}$ is bounded, by the Bolzano-Weierstrass Theorem 2.8 it has a convergent subsequence $\{x_{n_k}\}$ with limit, say, $x \in I$. The continuity of f implies that $f(x_{n_k}) \to f(x)$. Moreover, we also have

$$|x_{n_k} - w_{n_k}| < \frac{1}{n_k},$$

so $w_{n_k} \to x$ as well. The continuity of f then implies that $f(w_{n_k}) \to f(x)$. Thus, the sequence $\{f(x_{n_k}) - f(w_{n_k})\}$ converges to 0, which contradicts (4.3). So f must be uniformly continuous. $\qquad \square$

4.5.1   *Exercises*

(1) Let $a > 0$ be a real number. Consider the function $f : [a, \infty) \to \mathbb{R}$ defined as $f(x) = x^{-1}$. Prove that f is uniformly continuous.

(2) Consider the function $f(x) = x^2$ with domain $\mathbb{R}$. Prove that f is not uniformly continuous.

(3) Prove that the function $f : (0, 1) \to \mathbb{R}$ defined as $f(x) = x^{-2}$ is not uniformly continuous.

(4) Let $f, g : A \to \mathbb{R}$ be uniformly continuous functions, and let c be a real number.

  (a) Prove that cf is uniformly continuous.
  (b) Prove that $f + g$ is uniformly continuous.
  (c) Give an example in which fg is not uniformly continuous. In other words, the product of two uniformly continuous functions is not necessarily uniformly continuous.

(5) Give an example in which neither $f : A \to \mathbb{R}$ nor $g : A \to \mathbb{R}$ is uniformly continuous, but the product $fg : A \to \mathbb{R}$ is uniformly continuous.

(6) Let $f : A \to \mathbb{R}$ be uniformly continuous, and let $\{a_n\}$ be a Cauchy sequence in A.

  (a) Prove that $\{f(a_n)\}$ is a Cauchy sequence.
  (b) Show by an example that the previous part is false if f is only assumed to be continuous.

(7) Let $f : (a, b) \to \mathbb{R}$ be uniformly continuous. Show that f can be extended to a continuous function on $[a, b]$ as follows.

  (a) If $\{x_n\}$ is a convergent sequence in (a, b), prove that $\{f(x_n)\}$ is a convergent sequence.
  (b) Suppose that $\{x_n\}$ and $\{y_n\}$ are two sequences in (a, b) converging to a. Prove that the limits of $\{f(x_n)\}$ and $\{f(y_n)\}$ are equal.

(c) If $\{x_n\}$ is any sequence in (a, b) converging to a, define $f(a) = \lim f(x_n)$. Prove that $f(a)$ is well-defined and that the extended function f is continuous at a.

(d) Repeat the steps above for the other end point b to extend f to a continuous function on $[a, b]$.

The continuous function $f: [a, b] \to \mathbb{R}$ constructed above is called a **continuous extension** of the original function f.

(8) Give an example of a bounded continuous function $f: (a, b) \to \mathbb{R}$ that *cannot* be extended to a function that is continuous at a.

4.6 Monotone and Inverse Functions

For an integer $n \geq 2$, it is possible to show directly that the function $g(x) = x^{\frac{1}{n}}$ defined on $[0, \infty)$ is continuous. However, whether one uses the sequential definition or the ϵ-δ characterization of continuity, showing that g is continuous directly does involve quite a bit of work, especially if n is large. For example, try it for $n = 7$.

Notice that g is the inverse function of $f(x) = x^n$ defined on $[0, \infty)$. Since $h(x) = x$ is continuous on $\mathbb{R}$, it follows that f is continuous as well (Exercise (3) on page 93). This leads naturally to the following question. Is it possible to conclude from the continuity of f that its inverse function is continuous? If this is true, then it would save us a lot of work when dealing with inverse functions.

As you will see in this section, under some reasonable assumptions, the answer to the question above is yes. In particular, we can, in fact, conclude from the continuity of $f(x) = x^n$, with domain $[0, \infty)$, that its inverse function $g(x) = x^{\frac{1}{n}}$ is continuous on $[0, \infty)$. We begin with some relevant definitions.

4.6.1 *Monotone Functions*

Definition 4.8. Let $f: A \to \mathbb{R}$ be a function.

(1) We say that f is **increasing** if $a < b$ in A implies $f(a) \leq f(b)$.
(2) We say that f is **strictly increasing** if $a < b$ in A implies $f(a) < f(b)$.
(3) We say that f is **decreasing** if $a < b$ in A implies $f(a) \geq f(b)$.
(4) We say that f is **strictly decreasing** if $a < b$ in A implies $f(a) > f(b)$.
(5) We say that f is **monotone** if f is either increasing or decreasing.
(6) We say that f is **strictly monotone** if f is either strictly increasing or strictly decreasing.

Monotone functions will appear again in the next section when we discuss functions of bounded variation. Moreover, they are often used in applications in statistics, physics, and engineering. For example, cumulative distribution functions in probability theory are monotone functions. Furthermore, monotone functions defined on a closed bounded interval are Riemann integrable, as we will show in a later chapter.

Example 4.11. For a positive integer n, both $f(x) = x^n$ and $g(x) = x^{\frac{1}{n}}$ with domain $[0, \infty)$ are strictly increasing. The function x^2 with domain $[-1, 1]$ is neither increasing nor decreasing.

A function h is (strictly) increasing if and only if $-h$ is (strictly) decreasing. A strictly monotone function f is injective, so it has an inverse function f^{-1} ((1.1) on p. 6).

We want to show that if f is strictly monotone and continuous on an interval, then its inverse function is also strictly monotone and continuous. To prove this, we need the following result.

Theorem 4.10. *Let $f: I \to \mathbb{R}$ be a strictly monotone function on an interval I such that $f(I)$ is also an interval. Then f is continuous on I.*

Ideas of Proof. One can visualize this result as follows. If f is not continuous at some point $a \in I$, then since f is strictly monotone, there must be a "jump" in the graph of f at the point a. But this jump would violate the hypothesis that $f(I)$ is an interval. So f must be continuous on I.

Proof. We will consider the case when f is strictly increasing. The strictly decreasing case can be dealt with by considering $-f$. Pick a point $a \in I$. We must show that f is continuous at a. Let $\epsilon > 0$ be given. First consider the case when a is not an end point of I.

Since a is not an end point of I and f is strictly increasing, $f(a)$ is not an end point of the interval $f(I)$. So there exists $\delta_0 > 0$ with $\delta_0 \le \epsilon$ such that the closed interval $[f(a) - \delta_0, f(a) + \delta_0]$ is contained in $f(I)$. There exist unique elements α and β in I with $\alpha < a < \beta$ such that
$$f(\alpha) = f(a) - \delta_0 \quad \text{and} \quad f(\beta) = f(a) + \delta_0.$$
If we take $\delta = \min\{a - \alpha, \beta - a\}$, then
$$|x - a| < \delta \quad \text{implies} \quad \alpha \le a - \delta < x < a + \delta \le \beta.$$
Since f is strictly increasing, we have $f(\alpha) < f(x) < f(\beta)$. This implies that
$$f(a) - \epsilon \le f(a) - \delta_0 = f(\alpha) < f(x) < f(\beta) = f(a) + \delta_0 \le f(a) + \epsilon,$$
from which we conclude that
$$|f(x) - f(a)| < \epsilon.$$
This proves that f is continuous at a.

Next suppose a is the left end point of I, so $I = [a, b]$, $[a, b)$, or $[a, \infty)$. Then $f(a)$ is the left end point of $f(I)$ because f is strictly increasing. There exists $\delta_0 > 0$ with $\delta_0 \le \epsilon$ such that
$$[f(a), f(a) + \delta_0] \subseteq f(I).$$

So there exists a unique point $\beta \in I$ with

$$\beta > a \quad \text{and} \quad f(\beta) = f(a) + \delta_0.$$

If

$$|x - a| < \delta = \beta - a$$

with $x \in I$, then

$$a \leq x < \beta \quad \text{and} \quad f(a) \leq f(x) < f(\beta) \leq f(a) + \epsilon.$$

Thus, we have

$$0 \leq f(x) - f(a) < \epsilon,$$

showing that f is continuous at a. A similar argument can be used when a is the right end point of I. The details of this case is left to the reader as an exercise. $\square$

4.6.2 *Continuity of Inverse Functions*

We now use the previous theorem to show that the inverse function of a strictly monotone continuous function on an interval is also continuous.

Theorem 4.11. *Let f be a strictly monotone continuous function whose domain is an interval I. Then its inverse function f^{-1} is a strictly monotone continuous function on $f(I)$.*

Proof. Since f is strictly monotone, it is injective and has an inverse function f^{-1} with domain $f(I)$ and range I. Since f is continuous and injective, the set $f(I)$ is an interval (Exercise (4) on page 96). We first show that f^{-1} is also strictly monotone.

Suppose that f is strictly increasing. We will show that f^{-1} is also strictly increasing. Pick $y_0 < y_1$ in $f(I)$. Then there are unique points x_0 and x_1 in I such that

$$f(x_0) = y_0 \quad \text{and} \quad f(x_1) = y_1.$$

Since f is strictly increasing and $y_0 < y_1$, we have $x_0 < x_1$. By the definition of f^{-1}, we have

$$f^{-1}(y_0) = x_0 < x_1 = f^{-1}(y_1).$$

This shows that f^{-1} is also strictly increasing. A similar argument shows that if f is strictly decreasing, then so is f^{-1}.

Now f^{-1} is strictly monotone, whose domain is the interval $f(I)$. Moreover, we have $f^{-1}(f(I)) = I$, which is also an interval. Thus, Theorem 4.10 implies that f^{-1} is continuous on $f(I)$. $\square$

Example 4.12. For every positive integer n, the function $f(x) = x^n$ defined on $[0, \infty)$ is strictly increasing and continuous. Its range is also $[0, \infty)$. Theorem 4.11 tells us that its inverse function $f^{-1}(x) = x^{\frac{1}{n}}$ is also strictly increasing and continuous on $[0, \infty)$.

4.6.3   *Points of Discontinuity*

While a monotone function defined on an interval may not be continuous at all
the points of its domain, we now show that it does not have too many points of
discontinuity.

Theorem 4.12. *Let $f:(a,b) \to \mathbb{R}$ be a monotone function for some $a < b$. Then f
is discontinuous at at most countably many points in (a,b).*

Ideas of Proof. We show that, if f is not continuous at a point c, then the
monotonicity of f forces it to have a "jump" at c, inside of which we will pick
a rational number. The countability of $\mathbb{Q}$ will then be used to make the desired
conclusion.

Proof.   We will consider the case when f is increasing. The decreasing case can
be dealt with by considering $-f$. It suffices to show that the set of points of discontinuity of f is a subset of a countable set.

Suppose that f is discontinuous at a point c in (a,b). Since c is a limit point of
(a,b), we know that

$$f(c) \neq \lim_{x \to c} f. \qquad (4.4)$$

Since f is increasing, both

$$\alpha_c = \sup\{f(x) : c > x \in (a,b)\} \quad \text{and} \quad \beta_c = \inf\{f(x) : c < x \in (a,b)\}$$

are real numbers. We claim that

$$\alpha_c = \lim_{x \to c^-} f \quad \text{and} \quad \beta_c = \lim_{x \to c^+} f.$$

To prove this, we use the ϵ-δ characterization of one-sided limits (Exercise (12) on
page 90). Let $\epsilon > 0$ be given. There must exist $x_0 < c$ in (a,b) such that

$$\alpha_c - \epsilon < f(x_0) \leq \alpha_c.$$

Take $\delta = c - x_0$. Then

$$0 < c - x < \delta \quad \text{implies} \quad x_0 < x < c,$$

so we have

$$\alpha_c - \epsilon < f(x_0) \leq f(x) \leq \alpha_c.$$

So

$$|f(x) - \alpha_c| < \epsilon \quad \text{if} \quad 0 < c - x < \delta,$$

and we conclude that $\lim_{x \to c^-} f = \alpha_c$. A similar argument shows that $\lim_{x \to c^+} f = \beta_c$.
Now if both inequalities in

$$\lim_{x \to c^-} f = \alpha_c \leq f(c) \leq \beta_c = \lim_{x \to c^+} f$$

are equalities, then we would have $f(c) = \lim_{x \to c} f$ (Exercise (11) on page 90), contradicting (4.4). Thus, one of the above inequalities must be strict, so $\alpha_c < \beta_c$. Choose any rational number q_c in (α_c, β_c), which is possible by Theorem 1.3.

Starting at a point c of discontinuity of f, we have associated a rational number q_c to it. Moreover, if $d > c$ is another point of discontinuity of f, then the increasing hypothesis on f implies that $\beta_c \le \alpha_d$, so

$$q_c < \beta_c \le \alpha_d < q_d.$$

Since distinct points of discontinuity of f give rise to distinct rational numbers and the set $\mathbb{Q}$ of rational numbers is countable, the theorem is proved. $\square$

An interval (a, b) is an uncountable set. Theorem 4.12 says that a monotone function defined on (a, b) is continuous except possibly on a countable or finite subset.

There is a sort of converse to Theorem 4.12. Suppose that I is a bounded interval and that S is a finite or countable subset of I. Then there exists a monotone function f with domain I whose set D of points of discontinuity is exactly S. In fact, we can even insist that f be strictly increasing (or strictly decreasing, if one wishes). These issues are explored further in the exercises.

4.6.4 *Exercises*

(1) Consider the function $g(x) = x^{\frac{1}{n}}$ for some positive integer n.

    (a) Prove that $g(x)$ is uniformly continuous on $[0, 1]$.

    (b) Prove that $g(x)$ is uniformly continuous on $[1, \infty)$.

    (c) Using the previous two parts or otherwise, prove that $g(x)$ is uniformly continuous on $[0, \infty)$.

(2) For integers $n \ge 2$, prove that $f(x) = x^n$ is *not* uniformly continuous on $[0, \infty)$. Together with the previous exercise, this shows that uniform continuity is not preserved by the process of taking inverse functions.

(3) Suppose that $f : I \to \mathbb{R}$ is continuous on an interval $I = [a, b]$ and that f is injective.

    (a) If $f(a) < f(b)$, prove that f is strictly increasing.

    (b) If $f(a) > f(b)$, prove that f is strictly decreasing.

(4) Suppose that $f, g : A \to \mathbb{R}$ are increasing functions.

    (a) For a real number c, prove that cf is increasing if $c > 0$ and is decreasing if $c < 0$.

    (b) Prove that the sum $f + g$ is increasing.

(5) Suppose that $f, g : A \to \mathbb{R}$ are increasing functions.

    (a) Give an example in which the product fg is *not* monotone.

    (b) If, in addition, $f(x) > 0$ and $g(x) > 0$ for all $x \in A$, prove that fg is increasing.

(6) Suppose that there is a sequence of disjoint intervals I_n ($n \in \mathbb{Z}_+$) such that $\mathbb{R} = \bigcup_{n=1}^{\infty} I_n$. Let $f: \mathbb{R} \to \mathbb{R}$ be a function such that f is monotone on each I_n. Prove that f is continuous on $\mathbb{R}$, except possibly on a finite or countable set.

(7) Give an example of a bijective function $f: [0,1] \to [0,1]$ whose inverse function f^{-1} is *not* continuous at at least one point in $[0,1]$.

(8) Let $f: [a,b] \to \mathbb{R}$ be a monotone function. Prove that the set of points of discontinuity of f is either finite or countable.

(9) Let $f: (a,b) \to \mathbb{R}$ be an increasing function, and let c be a point in (a,b).

   (a) Prove that
$$\lim_{x \to c^+} f - \lim_{x \to c^-} f = \inf\{f(w) - f(x) : x < c < w, \, x, w \in I\}.$$

   (b) Prove that f is continuous at c if and only if
$$\inf\{f(w) - f(x) : x < c < w, \, x, w \in I\} = 0.$$

(10) Consider the **step function** $J: \mathbb{R} \to \mathbb{R}$ defined as
$$J(x) = \begin{cases} 0 & \text{if } x < 0, \\ 1 & \text{if } x \geq 0. \end{cases}$$

   (a) Prove that J is continuous on $\mathbb{R} \setminus \{0\}$ and discontinuous at 0.

   (b) Suppose that $x_1 < x_2 < \cdots < x_n$ are n distinct points in $\mathbb{R}$ and $a_1, \ldots, a_n$ are n positive real numbers. Sketch the graph of the function
$$f(x) = \sum_{k=1}^{n} a_k J(x - x_k). \qquad (4.5)$$

   (c) Prove that f is an increasing function.

   (d) Prove that f is continuous on $\mathbb{R} \setminus \{x_1, \ldots, x_n\}$ and discontinuous at $x_1, \ldots, x_n$.

   The function f is thus a monotone function with a prescribed finite set of points of discontinuity.

(11) Suppose that $x_1 < x_2 < \cdots < x_n$ are n distinct points in $\mathbb{R}$. Construct a strictly increasing function $f: \mathbb{R} \to \mathbb{R}$ that is continuous on $\mathbb{R} \setminus \{x_1, \ldots, x_n\}$ and discontinuous at $x_1, \ldots, x_n$.

4.7 Functions of Bounded Variation

We saw in the previous section that monotone functions have some nice properties. For example, a monotone function is continuous except possibly on a countable set. They are also closed under scalar multiplication, and the sum of two increasing functions is increasing (Exercise (4) on page 103). However, the difference and product of two monotone functions are not necessarily monotone (Exercise (5) on page 103).

For many purposes it is convenient to have a larger class of functions, containing all the monotone functions, that are closed under taking scalar multiples, sums, and products. In particular, such a class of functions must contain functions of the form $f = g - h$, where g and h are increasing. The class of differences of increasing functions, called *functions of bounded variation*, has the desired properties (see the exercises).

4.7.1 *Variation*

We will define functions of bounded variation in terms of the concept of *variation*, which measures how much the function fluctuates. Then we will show that they are equivalent to differences $g - h$ of increasing functions.

For the definition below, we need the concept of partition.

Definition 4.9. A **partition** of $[a, b]$ is a finite ordered set
$$P = \{a = x_0 < x_1 < \cdots < x_n = b\}.$$
The points $x_0, \ldots, x_n$ are called the **partition points** of P.

So a partition P divides the interval $[a, b]$ into n sub-intervals. Here n can be any positive integer, so the simplest partition of $[a, b]$ is $P = \{a < b\}$.

Definition 4.10. Let $f : [a, b] \to \mathbb{R}$ be a function, and let $[c, d] \subseteq [a, b]$ be a sub-interval.

(1) Suppose $P = \{x_0 < \cdots < x_n\}$ is a partition of $[c, d]$. The **variation of f with respect to P** is the sum
$$V(f; P) = \sum_{k=1}^{n} |f(x_k) - f(x_{k-1})|.$$

(2) The **variation of f on $[c, d]$** is defined as the extended real number
$$V(f; [c, d]) = \sup \{V(f; P) : P \text{ a partition of } [c, d]\}.$$
If $V(f; [c, d])$ is finite, then we say that f is of **bounded variation** on $[c, d]$. Otherwise, we take
$$V(f; [c, d]) = \infty.$$

The simplest functions of bounded variation are monotone functions.

Theorem 4.13. *Every monotone function on $[a, b]$ is of bounded variation.*

Proof. Suppose f is an increasing function on $[a, b]$. Let $P = \{x_0 < \cdots < x_n\}$ be any partition of $[a, b]$. Then we have a telescoping sum:
$$
\begin{aligned}
V(f; P) &= \sum_{k=1}^{n} |f(x_k) - f(x_{k-1})| \\
&= (f(x_1) - f(x_0)) + (f(x_2) - f(x_1)) + \cdots + (f(x_n) - f(x_{n-1})) \\
&= f(x_n) - f(x_0) = f(b) - f(a) < \infty.
\end{aligned}
$$

Since this is true for any partition P of $[a, b]$, we have shown that

$$V(f; [a, b]) = f(b) - f(a)$$

for an increasing function f, which is, therefore, of bounded variation. The case where f is decreasing is left as an exercise. □

There are, however, functions of bounded variation that are not monotone.

Example 4.13. Consider the function $f(x) = |x|$ defined on $[-1, 1]$, which is not monotone. We show that f is of bounded variation. With the partition $\{-1, 0, 1\}$, we see that

$$V(f; [-1, 1]) \geq |f(0) - f(-1)| + |f(1) - f(0)| = 2.$$

On the other hand, let $P = \{x_0 < \cdots < x_n\}$ be any partition of $[-1, 1]$. If one extra point w is added to the partition P to make a new partition Q, then

$$V(f; Q) \geq V(f; P).$$

Indeed, if $x_{k-1} < w < x_k$, then the Triangle Inequality gives

$$|f(x_k) - f(x_{k-1})| \leq |f(x_k) - f(w)| + |f(w) - f(x_{k-1})|.$$

This shows that $V(f; P) \leq V(f; Q)$.

Let Q be the partition obtained from P by addition the point 0, if it is not already in P. Since f is decreasing on $[-1, 0]$ and increasing on $[0, 1]$, the telescoping sum computation in Example 4.13 shows that

$$V(f; P) \leq V(f; Q) = (f(-1) - f(0)) + (f(1) - f(0)) = 2.$$

Since this is true for any partition P of $[-1, 1]$, we conclude that

$$V(f; [-1, 1]) \leq 2.$$

Together with the previous paragraph, we have $V(f, [-1, 1]) = 2$, so f is of bounded variation but not monotone.

4.7.2   Variations on Different Intervals

Theorem 4.14. *Suppose f is a function of bounded variation on $[a, b]$, and $[c, d] \subseteq [a, b]$. Then f is of bounded variation on $[c, d]$, and*

$$V(f; [c, d]) \leq V(f; [a, b]).$$

Proof.   Suppose that $P = \{x_0 < \cdots < x_n\}$ is a partition of $[c, d]$. Then

$$Q = \{a \leq x_0 < \cdots < x_n \leq b\}$$

is a partition of $[a, b]$, where a is inserted into P as the first point in Q if $a < c$, and b is inserted as the last point in Q if $d < b$. We have

$$V(f; P) \leq V(f; Q) \leq V(f; [a, b]).$$

Since this is true for any partition P of $[c, d]$, we conclude that

$$V(f; [c, d]) \leq V(f; [a, b]) < \infty,$$

as desired. □

The following result is a sort of converse to Theorem 4.14. It will also be used to characterize functions of bounded variation as differences of increasing functions.

Theorem 4.15. *Suppose $a < c < b$, and f is of bounded variation on $[a,c]$ and on $[c,b]$. Then*

$$V(f;[a,b]) = V(f;[a,c]) + V(f;[c,b]), \tag{4.6}$$

and f is of bounded variation on $[a,b]$.

Ideas of Proof. The main point is that it is possible to add a point to a given partition, as in Example 4.13. Moreover, the desired equality holds for a fixed partition of $[a,b]$ that includes c.

Proof. The assertion that f is of bounded variation on $[a,b]$ will follow from the equality (4.6), since $V(f;[a,c])$ and $V(f;[c,b])$ are assumed to be finite. To prove (4.6), let $P = \{x_0 < \cdots < x_n\}$ be any partition of $[a,b]$. The following argument is similar to the one used in Example 4.13. Let Q be the partition of $[a,b]$ obtained from P by adding the point c, if it is not already in P. Then Q gives rise to the partitions

$$Q_1 = \{a = x_0 < \cdots < c\} \quad \text{and} \quad Q_2 = \{c < \cdots < x_n = b\}$$

of $[a,c]$ and $[c,b]$, respectively. As in Example 4.13, we have

$$V(f;P) \le V(f;Q) = V(f;Q_1) + V(f;Q_2) \le V(f;[a,c]) + V(f;[c,b]).$$

Since this is true for any partition P of $[a,b]$, we infer that

$$V(f;[a,b]) \le V(f;[a,c]) + V(f;[c,b]) < \infty. \tag{4.7}$$

It remains to prove the reverse inequality.

Pick any partitions

$$P_1 = \{x_0 < \cdots < x_r = c\} \quad \text{and} \quad P_2 = \{c = x_r < \cdots < x_{r+s}\}$$

of $[a,c]$ and $[c,b]$, respectively. Splicing these partitions together, we obtain a partition

$$P = \{x_0 < \cdots < x_r < \cdots < x_{r+s}\}$$

of $[a,b]$ with $x_r = c$. Then we have

$$V(f;P_1) + V(f;P_2) = V(f;P) \le V(f;[a,b]).$$

Since this is true for any partition P_1 of $[a,c]$, we have

$$V(f;[a,c]) + V(f;P_2) \le V(f;[a,b]).$$

Likewise, this is true for any partition P_2 of $[c,b]$, so

$$V(f;[a,c]) + V(f;[c,b]) \le V(f;[a,b]).$$

Combining this inequality with (4.7), we conclude that (4.6) is true. $\qquad\square$

4.7.3   *Characterization*

Now we are ready to show that every function of bounded variation can be written as the difference of two increasing functions.

Theorem 4.16. *Suppose f is a function of bounded variation on $[a, b]$. Then there exist increasing functions $g, h : [a, b] \to \mathbb{R}$ such that $f = g - h$.*

Proof.   Consider the function $g : [a, b] \to \mathbb{R}$ defined as

$$g(x) = \begin{cases} 0 & \text{if } x = a, \\ V(f; [a, x]) & \text{if } a < x \le b. \end{cases} \tag{4.8}$$

This function g is well-defined by Theorem 4.14. First we observe that g is increasing. For $a \le c < d \le b$, Theorem 4.15 implies that

$$g(d) = V(f; [a, d]) = g(c) + V(f; [c, d]),$$

so

$$g(d) - g(c) = V(f; [c, d]) \ge 0.$$

Thus, g is increasing. We define h as the difference $h = g - f$. Then $f = g - h$, so it remains to show that h is increasing.

For $a \le c < d \le b$, using the trivial partition $\{c < d\}$ of $[c, d]$, we have

$$g(d) - g(c) = V(f; [c, d]) \ge |f(d) - f(c)| \ge f(d) - f(c).$$

So we have

$$h(d) = g(d) - f(d) \ge g(c) - f(c) = h(c),$$

and h is, therefore, increasing.   □

Theorem 4.16 allows us to carry certain nice properties of monotone functions to functions of bounded variation. For example, since monotone functions are Riemann integrable, so are functions of bounded variation. The following result gives another example.

Corollary 4.2. *Suppose f is a function of bounded variation on $[a, b]$. Then the set of points of discontinuity of f is either finite or countable.*

Proof.   Write f as $g - h$, where g and h are increasing, hence monotone, functions. Each of g and h has at most countably many points of discontinuity (Theorem 4.12). If f is discontinuous at a point c, then at least one of g and h is discontinuous at c. So the set of points of discontinuity of f is a subset of the union of two finite or countable sets, which is at most countable.   □

The converse of Theorem 4.16 is also true. In other words, if $g, h : [a, b] \to \mathbb{R}$ are increasing functions, then their difference $f = g - h$ is of bounded variation on $[a, b]$ (Exercise (4) below). Thus, functions of bounded variation are exactly differences of increasing functions. Some nice properties of the class of functions of bounded variation are explored in the exercises.

4.7.4 *Exercises*

(1) Let $f:[a,b] \to \mathbb{R}$ be of bounded variation on $[a,b]$. Prove that

$$|f(x)| \leq |f(a)| + V(f;[a,b])$$

for all x in $[a,b]$. In particular, f is bounded on $[a,b]$.

(2) Finish the proof of Theorem 4.13 by proving the decreasing case.

(3) Let $f,g:[a,b] \to \mathbb{R}$ be of bounded variation on $[a,b]$.

   (a) For a real number c, prove that cf is of bounded variation on $[a,b]$.

   (b) Prove that $f+g$ is of bounded variation on $[a,b]$.

   (c) Prove that $f-g$ is of bounded variation on $[a,b]$.

   (d) Prove that the product fg is of bounded variation on $[a,b]$.

   These properties imply that functions of bounded variation on $[a,b]$ form an **algebra of functions**. In general, an algebra of functions is a set of functions that is closed under scalar multiplication, addition, and product.

(4) If $g,h:[a,b] \to \mathbb{R}$ are increasing functions, prove that their difference $f = g - h$ is of bounded variation on $[a,b]$.

(5) Let $f:[a,b] \to \mathbb{R}$ be of bounded variation on $[a,b]$. Suppose that there exists a real number $r > 0$ such that $f(x) \geq r$ for all $x \in [a,b]$. Prove that $\frac{1}{f}$ is of bounded variation on $[a,b]$.

(6) In each case, compute the variation $V(f,[a,b])$.

   (a) $f(x) = \sin(x)$ and $[a,b] = [0,2\pi]$.

   (b) $f(x) = \cos^2(x)$ and $[a,b] = [0,\pi]$.

   (c) $f(x) = x^2 - x$ and $[a,b] = [-2,2]$.

(7) Consider the function f defined on $[0,\frac{1}{\pi}]$ as

$$f(x) = \begin{cases} \sin\left(\frac{1}{x}\right) & \text{if } x \neq 0, \\ 0 & \text{if } x = 0. \end{cases}$$

   (a) Prove that f is bounded on $[0,\frac{1}{\pi}]$.

   (b) Prove that f is not of bounded variation on $[0,\frac{1}{\pi}]$. So a bounded function is not necessarily of bounded variation.

   (c) If $0 < a < \frac{1}{\pi}$, prove that $f(x) = \sin(\frac{1}{x})$ is of bounded variation on $[a,\frac{1}{\pi}]$.

(8) Let $f:[a,b] \to \mathbb{R}$ be of bounded variation on $[a,b]$.

   (a) Prove that $\lim_{x \to c^-} f$ exists when $a < c \leq b$.

   (b) Prove that $\lim_{x \to c^+} f$ exists when $a \leq c < b$.

(9) Let $f:[a,b] \to \mathbb{R}$ be of bounded variation on $[a,b]$. Suppose that f is continuous on (a,b). Prove that f is uniformly continuous on (a,b).

(10) Let $f:[a,b] \to \mathbb{R}$ be a function. Suppose that $P = \{x_0 < \cdots < x_n\}$ is a partition of $[a,b]$ such that f is monotone on each subinterval $[x_{k-1},x_k]$.

   (a) Compute $V(f;[a,b])$ in terms of $f(x_k)$ for $k = 0,1,\ldots,n$.

   (b) Prove that f is of bounded variation on $[a,b]$.

(11) Prove that the characteristic function $\chi_\mathbb{Q}$ (4.2) is not of bounded variation on $[a, b]$.

(12) Let $f:[a, b] \to \mathbb{R}$ be a function, and let P be a partition of $[a, b]$. Suppose that Q is another partition of $[a, b]$ such that all the points in P are also in Q. Prove that $V(f; P) \le V(f; Q)$.

(13) Let $f:[a, b] \to \mathbb{R}$ be of bounded variation on $[a, b]$, and let g be defined as in (4.8).

(a) Prove that $g(x) \ge 0$ for all $x \in [a, b]$.

(b) Prove that $V(f; [a, b]) = V(g; [a, b])$.

(c) If g is continuous at $c \in (a, b)$, prove that f is also continuous at c.

(14) Let $f:[a, b] \to \mathbb{R}$ be of bounded variation on $[a, b]$. Prove that there exist strictly increasing functions $f_1, f_2:[a, b] \to \mathbb{R}$ such that $f = f_1 - f_2$.

(15) Let r be a real number with $0 < r < 1$. Consider the function $g:[0, 1] \to \mathbb{R}$ defined as

$$g(x) = \begin{cases} -r^n & \text{if } x = \frac{1}{n} \text{ with } n \text{ odd}, \\ r^n & \text{if } x = \frac{1}{n} \text{ with } n \text{ even}, \\ 0 & \text{otherwise}. \end{cases}$$

Prove that g is of bounded variation.

(16) Consider the function $h:[0, 1] \to \mathbb{R}$ defined as

$$h(x) = \begin{cases} 1 & \text{if } x = \frac{1}{n} \text{ with } n \text{ even}, \\ 0 & \text{otherwise}. \end{cases}$$

Prove that h is not of bounded variation.

(17) Show by an example that the composition $f \circ g$ of two functions f and g of bounded variation is not necessarily of bounded variation.

4.8   Additional Exercises

(1) Let $f:[a, \infty) \to \mathbb{R}$ be a function, and let L be a real number. We write $\lim_{x \to \infty} f = L$ if for every sequence $a_n \to \infty$ with each $a_n > a$, we have $f(a_n) \to L$. In this case, we say that the **limit of f at ∞ is L**. Prove that $\lim_{x \to \infty} f = L$ if and only if for every $\epsilon > 0$, there exists $M > 0$ with $M > a$ such that $x > M$ implies $|f(x) - L| < \epsilon$.

(2) Using the previous exercise as a guide:

(a) Write down a reasonable definition of $\lim_{x \to -\infty} f = L$ in terms of sequences.

(b) Prove an ϵ-M characterization of $\lim_{x \to -\infty} f = L$.

(3) Let $f:[a, \infty) \to \mathbb{R}$ be a continuous function such that $\lim_{x \to \infty} f$ exists. Prove that f is uniformly continuous on $[a, \infty)$.

(4) Let $f:[a, \infty) \to \mathbb{R}$ be a function. We write $\lim_{x \to \infty} f = \infty$ if for every sequence $a_n \to \infty$ with each $a_n > a$, we have $f(a_n) \to \infty$. Prove that $\lim_{x \to \infty} f = \infty$ if and only if for every $\alpha > 0$, there exists $M > 0$ with $M > a$ such that $x > M$ implies $f(x) > \alpha$.

(5) Using the previous exercise as a guide:

    (a) Write down a reasonable definition of $\lim_{x \to \infty} f = -\infty$ in terms of sequences. Then prove an α-M characterization of $\lim_{x \to \infty} f = -\infty$.

    (b) Repeat (a) for $\lim_{x \to -\infty} f = \infty$.

    (c) Repeat (a) for $\lim_{x \to -\infty} f = -\infty$.

(6) Let $f:[a, b] \to \mathbb{R}$ be a monotone function. Prove that every subinterval $[c, d] \subseteq [a, b]$ contains uncountably many points at which f is continuous.

(7) Suppose that $a \in A \subseteq \mathbb{R}$. We say that a is an **isolated point** of A if there exists an open interval J such that $J \cap A = \{a\}$.

    (a) Prove that a is an isolated point of A if and only if it is not a limit point of A.

    (b) Let $f:A \to \mathbb{R}$ be a function. Prove that f is continuous at every isolated point of A.

(8) Let f be a polynomial with odd degree. Prove that there exists $a \in \mathbb{R}$ such that $f(a) = 0$.

(9) Let $f, g: A \to \mathbb{R}$ be two functions. Define $h: A \to \mathbb{R}$ as

$$h(x) = \max\{f(x), g(x)\} = \begin{cases} f(x) & \text{if } f(x) \geq g(x), \\ g(x) & \text{if } f(x) < g(x). \end{cases}$$

    (a) If f and g are both continuous at $a \in A$, prove that h is continuous at a.

    (b) Prove that $F(x) = \max\{f(x), 0\}$ is continuous at any point at which f is continuous.

(10) Let $f:[0, 2] \to \mathbb{R}$ be a continuous function such that $f(0) = f(2)$. Prove that there exists $a \in [0, 1]$ such that $f(a) = f(a + 1)$.

(11) Let $f:[a, b] \to \mathbb{R}$ be a continuous function. Suppose that the set $f([a, b])$ is either finite or countable. Prove that f is a constant function, i.e., $f(x) = c$ for all $x \in [a, b]$ for some fixed real number c.

(12) Let $f:A \to \mathbb{R}$ be a uniformly continuous function on a bounded set A.

    (a) Prove that f is bounded on A.

    (b) Show by examples that f may not be bounded on A if it is merely assumed to be continuous or if A is not bounded.

(13) A function $f: \mathbb{R} \to \mathbb{R}$ is said to be p-**periodic** if there exists $p > 0$ such that $f(x + p) = f(x)$ for all $x \in \mathbb{R}$. For example, $\sin(x)$ is 2π-periodic with $p = 2\pi$. Prove that a continuous periodic function is uniformly continuous and bounded on $\mathbb{R}$.

(14) Let $f:[a, b] \to \mathbb{R}$ be a continuous function such that $f(a) = f(b)$. Prove that there exists a *periodic* uniformly continuous function $g: \mathbb{R} \to \mathbb{R}$ such that $g(x) =$

$f(x)$ for $x \in [a, b]$. Such a function g is called a **periodic extension** of f to $\mathbb{R}$.

(15) A function $f: A \to \mathbb{R}$ is called a **Lipschitz function** if there exists a real number $M > 0$ such that

$$|f(a) - f(b)| \le M|a - b|$$

for all $a, b \in A$.

(a) Prove that a Lipschitz function is uniformly continuous.
(b) Prove that $f(x) = \sqrt{x} : [c, \infty) \to \mathbb{R}$ for any $c > 0$ is a Lipschitz function.
(c) Give an example of a uniformly continuous function that is not a Lipschitz function.

(16) In each case, determine whether the function $f: A \to \mathbb{R}$ is a Lipschitz function.

(a) $f(x) = x^2$ and $A = [a, b]$.
(b) $f(x) = x^2$ and $A = [0, \infty)$.
(c) $f(x) = \frac{1}{x}$ and $A = [1, b]$ for some $b > 1$.
(d) $f(x) = \frac{1}{1+x^2}$ and $A = \mathbb{R}$.

(17) Let $f, g: A \to \mathbb{R}$ be Lipschitz functions.

(a) Prove that $f + g: A \to \mathbb{R}$ is a Lipschitz function.
(b) Suppose, in addition, that f and g are bounded on A. Prove that the product $fg: A \to \mathbb{R}$ is a Lipschitz function.

(18) Consider the function $f: (0, \infty) \to \mathbb{R}$ defined as

$$f(x) = \begin{cases} \frac{1}{q} & \text{if } x \in \mathbb{Q} \text{ with } x = \frac{p}{q} \text{ in lowest terms,} \\ 0 & \text{if } x \notin \mathbb{Q}. \end{cases}$$

(a) Prove that f is discontinuous at every rational $x \in (0, \infty)$.
(b) Prove that f is continuous at every point $x \notin \mathbb{Q}$ with $x \in (0, \infty)$.

This is called the **Thomae function**.

(19) Consider the function $f: [0, 1] \to [0, 1]$ defined as

$$f(x) = \begin{cases} x & \text{if } x \in \mathbb{Q}, \\ 1 - x & \text{if } x \notin \mathbb{Q}. \end{cases}$$

(a) Prove that f is a bijection.
(b) Prove that f is not monotone on any subinterval of $[0, 1]$.
(c) Prove that f is continuous at $\frac{1}{2}$ and discontinuous on $[0, \frac{1}{2}) \cup (\frac{1}{2}, 1]$.

This exercise illustrates that a function defined on a closed bounded interval may satisfy the conclusion of the Intermediate Value Theorem without being continuous on any subinterval.

(20) Let $f: A \to \mathbb{R}$ be a function, and suppose that $a \in A$. We say that f is **left continuous at** a if for every sequence $a_n \to a$ with each $a_n \in A$ and $a_n < a$, the sequence $\{f(a_n)\}$ converges to $f(a)$.

(a) Prove that f is left continuous at a if and only if for every $\epsilon > 0$, there exists $\delta > 0$ such that $0 \le a - x < \delta$ with $x \in A$ implies $|f(x) - f(a)| < \epsilon$.
(b) Give a reasonable definition of **right continuous at** a.
(c) Prove an ϵ-δ characterization of right continuity.
(d) Prove that f is continuous at a if and only if f is left continuous at a and right continuous at a.

(21) Give an example in which a function f is left continuous at a point a but is not right continuous at a.

(22) Consider the function $f:[1,3] \to \mathbb{R}$ defined as

$$f(x) = \begin{cases} \left(\dfrac{4x+1}{14x-3}\right)^{\frac{1}{2}} & \text{if } 1 \le x < 2, \\ 3x - 2 & \text{if } 2 \le x \le 3. \end{cases}$$

Determine if f is left continuous at 2, right continuous at 2, or neither.

(23) Let $\{x_n\}$ be a strictly increasing convergent sequence, and let $\sum a_n$ be a convergent series with each $a_n > 0$. Define a function $f:\mathbb{R} \to \mathbb{R}$ as

$$f(x) = \sum_{n=1}^{\infty} a_n J(x - x_n),$$

where $J(x)$ is the step function from Exercise (10) on p. 104.

(a) Prove that f is a well-defined increasing function.
(b) Prove that $f(x) = 0$ for $x < x_1$ and $f(x) = \sum a_n$ for $x \ge \lim x_n$.
(c) Prove that f is continuous on $\mathbb{R} \setminus \{x_n : n \in \mathbb{Z}_+\}$.
(d) Prove that f is discontinuous at each x_n.

(24) A subset $A \subseteq \mathbb{R}$ is **open** if for each $a \in A$, there exists $\delta > 0$ such that the open interval $(a - \delta, a + \delta)$ is a subset of A. A subset $A \subseteq \mathbb{R}$ is **closed** if $\mathbb{R} \setminus A$ is open. These concepts were introduced in Exercises (14) – (18) on p. 26. Prove that $A \subseteq \mathbb{R}$ is closed if and only if A contains all of its limit points.

(25) A subset $S \subseteq A$ is said to be **open in** A if there exists an open set O such that $S = A \cap O$. Let $f:A \to \mathbb{R}$ be a function.

(a) Prove that f is continuous on A if and only if for every open set $U \subseteq \mathbb{R}$, the set $f^{-1}(U)$ is open in A.
(b) If $f:(a,b) \to \mathbb{R}$ is continuous and strictly monotone, prove that $f(U)$ is open for every open U in (a,b).
(c) Show by an example that the conclusion of the previous part is not necessarily true if f is not required to be strictly monotone.

(26) In this exercise, you will show that a subset $A \subseteq \mathbb{R}$ is open if and only if it is the disjoint union of at most countably many open intervals.

(a) Prove that a disjoint union of finitely or countably many open intervals is open.

(b) Let $A \subseteq \mathbb{R}$ be open and non-empty. For each $a \in A$, consider the set $I_a = \bigcup J$, where $J \subseteq A$ is an open interval containing a. Prove that I_a is an open interval containing a with $I_a \subseteq A$.

(c) If A is open and $a, b \in A$, prove that $I_a \cap I_b \neq \varnothing$ implies $I_a = I_b$.

(d) Using the previous two parts, prove that if A is open, then A is the disjoint union of at most countably many open intervals.

(27) Let $f: A \to \mathbb{R}$ be a continuous and bounded function on a closed set A. In this exercise, you will prove that there exists a continuous function $g: \mathbb{R} \to \mathbb{R}$ such that $g(x) = f(x)$ for all $x \in A$. Since $\mathbb{R} \setminus A$ is open, by the previous exercise, it is the disjoint union of at most countably many open intervals $I_k = (a_k, b_k)$ (or $(-\infty, b_k)$ or (a_k, ∞)) with end points in A. Define $g: \mathbb{R} \to \mathbb{R}$ as

$$
g(x) = \begin{cases}
f(x) & \text{if } x \in A, \\
f(a_k) + (x - a_k)\dfrac{f(b_k) - f(a_k)}{b_k - a_k} & \text{if } x \in I_k = (a_k, b_k), \\
f(b_k) & \text{if } x \in I_k = (-\infty, b_k), \\
f(a_k) & \text{if } x \in I_k = (a_k, \infty).
\end{cases}
$$

Prove that g is continuous on $\mathbb{R}$. This is a special case of the **Tietze Extension Theorem**. The function g is called a **continuous extension** of f to $\mathbb{R}$.

Chapter 5

Differentiation

The derivative of a function measures its rate of change. In terms of its graph, the derivative of a nice function can be interpreted as the slope of the graph. It is a very important concept in analysis and in the sciences. Basic rules regarding the computation of derivatives are discussed in section 5.1. In section 5.2 we discuss the Mean Value Theorem, which roughly says that the slope of any secant line of the graph of a function can be achieved as the derivative at some point. Several important results are consequences of the Mean Value Theorem, including Taylor's Theorem, which is discussed in section 5.3, and the Fundamental Theorems of Calculus.

5.1   The Derivative

We assume that the reader is familiar with the interpretations of the derivative as a rate of change or, geometrically, as the slope of the tangent line to the graph. Thus, we will concentrate on proving properties and consequences of the derivative in details. We begin with its definition. Recall that an open interval is an interval of the form (a, b), $(-\infty, b)$, (a, ∞), or $(-\infty, \infty)$.

5.1.1   *Definition of Derivative*

Definition 5.1. Let $f: I \to \mathbb{R}$ be defined on an open interval I, and let $c \in I$. The **derivative of** f **at** c is defined as the limit

$$f'(c) = \lim_{x \to c} \frac{f(x) - f(c)}{x - c},$$

if it exists. If this is the case, we say that f is **differentiable at** c. If this limit does not exist, then we say that f is not differentiable at c. If f is differentiable at every point in I, then we say that f is **differentiable on** I and write f' for the function whose value at $x \in I$ is $f'(x)$. Similarly, write

$$f^{(n)}(c) = (f^{(n-1)})'(c),$$

if the derivative exists, and call it the nth **derivative** of f at c. The second and the third derivatives are also written as f'' and f''', respectively, if they exist. The 0th derivative $f^{(0)}$ is defined as f itself. We say that f is **continuously differentiable** if f' exists and is continuous. If $f^{(n)}$ exists for all n, then f is called **infinitely differentiable**.

A function f is *not* differentiable at c if and only if there exists a sequence $\{x_n\} \in I \setminus \{c\}$ converging to c such that the sequence

$$\left\{ \frac{f(x_n) - f(c)}{x_n - c} \right\}$$

does not converge. From now on, when we say f is differentiable at c, we automatically assume that f is defined on some open interval containing c. The alternative notation $\frac{df}{dx}$ is also used for $f'(x)$.

The limit of a function was defined in Definition 4.2 in terms of sequences. An ϵ-δ characterization of limit was given in Theorem 4.3. As an exercise, the reader should write down in details these two equivalent formulations of the derivative.

5.1.2 *Differentiability and Continuity*

The first important property of differentiability is that it implies continuity.

Theorem 5.1. *Let $f : I \to \mathbb{R}$ be defined on an open interval I, and let $c \in I$. If f is differentiable at c, then f is continuous at c.*

Proof. Let $\{x_n\}$ be a sequence in $I \setminus \{c\}$ with $\lim x_n \to c$. Since f is differentiable at c, we have

$$\lim_{n \to \infty} \frac{f(x_n) - f(c)}{x_n - c} = f'(c).$$

Therefore, we have

$$f(x_n) - f(c) = \left(\frac{f(x_n) - f(c)}{x_n - c} \right) (x_n - c) \to f'(c) \cdot 0 = 0,$$

i.e., $\lim f(x_n) = f(c)$. So f is continuous at c. □

The converse of the above theorem is not true, as the following example illustrates.

Example 5.1. Consider the function $f(x) = |x|$ defined on $\mathbb{R}$. This function is continuous on $\mathbb{R}$. However, it is not differentiable at 0. Indeed, let $x_n = \frac{1}{n}$ if $n > 1$ is even and $x_n = -\frac{1}{n}$ if $n \geq 1$ is odd. Then $x_n \to 0$, and $f(x_n) = \frac{1}{n}$ for all n. So we have

$$\frac{f(x_n) - f(0)}{x_n - 0} = \begin{cases} 1 & \text{if } n \text{ is even,} \\ -1 & \text{if } n \text{ is odd.} \end{cases}$$

This shows that $\lim \frac{f(x_n) - f(0)}{x_n - 0}$ does not exist, and f is not differentiable at 0.

What makes the function $f(x) = |x|$ not differentiable at 0 is the corner on the graph of f at $x = 0$. Using a combination of functions with more and more corners on their graphs, it is, in fact, possible to construct a function g that is continuous everywhere but is nowhere differentiable. We will present such an example in a later chapter when we have the machinery of series of functions at our disposal.

Be careful that Theorem 5.1 does *not* say that f' is continuous at c. It also does not say that f is uniformly continuous, even if f is differentiable on I. These issues are explored further in the exercises.

5.1.3   *Arithmetics of Derivatives*

As in the case of sequences, it is convenient to know how derivative behaves with respect to arithmetic operations.

Theorem 5.2. *Let $f, g: I \to \mathbb{R}$ be defined on an open interval I, and let $c \in I$. Suppose that both f and g are differentiable at c. Then we have:*

(1) $(af)'(c) = af'(c)$ *for every $a \in \mathbb{R}$.*
(2) $(f + g)'(c) = f'(c) + g'(c)$.
(3) **Product Rule:** $(fg)'(c) = f(c)g'(c) + f'(c)g(c)$.
(4) **Quotient Rule:** *If $g(c) \neq 0$, then*

$$\left(\frac{f}{g}\right)'(c) = \frac{f'(c)g(c) - f(c)g'(c)}{g(c)^2}.$$

In the first case, when we write $(af)'(c) = af'(c)$, we mean that the function af is differentiable at c and that its derivative at c is $af'(c)$. Similar remarks apply to the other cases.

Proof.   We will prove the last case, which is the most difficult of the four cases, and leave the first three cases to the reader as an exercise. So suppose that $g(c) \neq 0$. Since g is also continuous at c (Theorem 5.1), there exists an open interval $J \subseteq I$ containing c such that $g(x) \neq 0$ for each $x \in J$. Pick any sequence $\{x_n\}$ in $I \smallsetminus \{c\}$ with $x_n \to c$. Then there exists a positive integer N such that $n \geq N$ implies $x_n \in J$, $g(x_n) \neq 0$, and

$$
\begin{aligned}
\frac{\left(\frac{f}{g}\right)(x_n) - \left(\frac{f}{g}\right)(c)}{x_n - c} &= \frac{f(x_n)g(c) - f(c)g(x_n)}{(x_n - c)g(x_n)g(c)} \\
&= \frac{(f(x_n) - f(c))\,g(c) - f(c)\,(g(x_n) - g(c))}{(x_n - c)g(x_n)g(c)} \\
&= \left(\frac{f(x_n) - f(c)}{x_n - c}\right)\frac{1}{g(x_n)} - \left(\frac{f(c)}{g(x_n)g(c)}\right)\left(\frac{g(x_n) - g(c)}{x_n - c}\right) \\
&\to \frac{f'(c)}{g(c)} - \frac{f(c)g'(c)}{g(c)^2}.
\end{aligned}
$$

In the last step, we used the differentiability of f and g at c and the continuity of g at c. We obtain the desired expression for $(\frac{f}{g})'(c)$ when we write the above difference as one fraction.   □

Example 5.2. Consider the function $f(x) = x^n$ defined on $\mathbb{R}$, where n is a positive integer. We show by induction that f is differentiable on $\mathbb{R}$ and that

$$f'(c) = nc^{n-1}$$

for any real number c. If $n = 1$, then

$$f(x) - f(c) = x - c,$$

so $f'(c) = \lim_{x \to c} 1 = 1c^0$. Suppose that $(x^n)'(c) = nc^{n-1}$ for some $n \geq 1$. Then

$$(x^{n+1})'(c) = (x^n \cdot x)'(c) = (x^n)(c) \cdot 1 + (x^n)'(c) \cdot (c)$$
$$= c^n + nc^{n-1} \cdot c = (n+1)c^n.$$

Thus, by induction we have shown that $f'(x) = nx^{n-1}$ for $f(x) = x^n$, where n is any positive integer and $x \in \mathbb{R}$.

5.1.4 Chain Rule

The next derivative rule is about the composition of two functions. To motivate it, consider the function $h(x) = \sqrt{1 - x^2}$. Computing its derivative using the definition is quite inconvenient. However, note that h is the composition $g \circ f$, where the inside function is $f(x) = 1 - x^2$ and the outside function is $\sqrt{x}$. It is not difficult at all to compute the derivatives of f and g using the definition. The reader should give it a try. This leads naturally to the question: Is there a way to express the derivative of a composition in terms of the individual functions and their derivatives? The answer is yes, and that is the content of the following result.

Theorem 5.3 (Chain Rule). *If f is differentiable at c and g is differentiable at $f(c)$, then the composition $g \circ f$ is differentiable at c and*

$$(g \circ f)'(c) = g'(f(c)) \cdot f'(c).$$

Ideas of Proof. The plan is to write down two suitable functions whose limits as $x \to c$ are $g'(f(c))$ and $f'(c)$, respectively.

Proof. Suppose I is an open interval in $Dom(f)$ containing c, and define the function φ by

$$\varphi(y) = \begin{cases} \dfrac{g(y) - g(f(c))}{y - f(c)} & \text{if } y \in Dom(g) \text{ and } y \neq f(c), \\ g'(f(c)) & \text{if } y = f(c). \end{cases}$$

The differentiability of g at $f(c)$ means

$$\varphi(f(c)) = g'(f(c)) = \lim_{y \to f(c)} \frac{g(y) - g(f(c))}{y - f(c)} = \lim_{y \to f(c)} \varphi(y),$$

which in turn means φ is continuous at $f(c)$. From the definition of φ, we have

$$\frac{g(f(x)) - g(f(c))}{x - c} = \varphi(f(x)) \cdot \frac{f(x) - f(c)}{x - c} \qquad (5.1)$$

for $x \in Dom(g \circ f)$ with $x \neq c$. Note that $(g \circ f)'(c)$ is the limit as $x \to c$ of the left-hand side of (5.1). Moreover, the limit as $x \to c$ of the second factor on the right-hand side of (5.1) is $f'(c)$. Since f is also continuous at c, we have $\lim_{x \to c} f(x) = f(c)$. The continuity of φ at $f(c)$ now implies

$$\lim_{x \to c} \varphi(f(x)) = \varphi(f(c)) = g'(f(c)),$$

which proves the theorem. $\qquad \square$

Example 5.3. Using the limit definition, one finds that if $f(x) = 1 - x^2$ and $g(x) = \sqrt{x}$, then $f'(x) = -2x$ and $g'(x) = \frac{1}{2\sqrt{x}}$ for $x > 0$. Therefore, chain rule says that the derivative of $h(x) = \sqrt{1 - x^2} = (g \circ f)(x)$ is

$$h'(x) = g'(f(x)) \cdot f'(x) = \frac{-2x}{2\sqrt{1 - x^2}}$$

for $-1 < x < 1$.

5.1.5 *Derivatives of Inverse Functions*

In Theorem 4.11 we saw that a strictly monotone continuous function defined on an interval has a strictly monotone continuous inverse function. The following result says that if the function is differentiable with non-zero derivative, then its inverse function is also differentiable.

Theorem 5.4 (Inverse Function Theorem). *Let f be a strictly monotone continuous function defined on an open interval I, c be a point in I, and f be differentiable at c with $f'(c) \neq 0$. Then its inverse function $g = f^{-1}$ is differentiable at $d = f(c)$ and*

$$g'(d) = \frac{1}{f'(c)}.$$

Ideas of Proof. The main point is that $\frac{1}{f'(c)}$ can be computed as the limit of the reciprocal of the difference quotient that appears in the definition of $f'(c)$.

Proof. The domain $J = f(I)$ of g is an open interval (Exercise (4) on page 96). Given $\epsilon > 0$ we need to show that there exists $\delta > 0$ such that $0 < |y - d| < \delta$ implies

$$\left| \frac{g(y) - g(d)}{y - d} - \frac{1}{f'(c)} \right| = \left| \frac{g(y) - c}{f(g(y)) - f(c)} - \frac{1}{f'(c)} \right| < \epsilon. \qquad (5.2)$$

The assumptions that f is strictly monotone and that $f'(c) \neq 0$ imply the equality

$$\frac{1}{f'(c)} = \lim_{x \to c} \frac{x - c}{f(x) - f(c)}$$

by Exercise (2d) on page 89. Therefore, with the given ϵ, there exists $\gamma > 0$ such that $0 < |x - c| < \gamma$ implies

$$\left| \frac{x - c}{f(x) - f(c)} - \frac{1}{f'(c)} \right| < \epsilon. \tag{5.3}$$

Since g is continuous on J (Theorem 4.11), there exists $\delta > 0$ such that $0 < |y - d| < \delta$ implies

$$0 < |g(y) - g(d)| = |g(y) - c| < \gamma.$$

The previous inequality implies (5.3) with $x = g(y)$, which is the desired inequality (5.2). $\qquad\square$

The derivative formula for the inverse function can be derived using chain rule. In fact, since $x = g(f(x))$, differentiating both sides we obtain $1 = g'(f(x)) \cdot f'(x)$. Writing $y = f(x)$, we then obtain

$$g'(y) = \frac{1}{f'(x)}.$$

However, the reader is cautioned that this line of reasoning does *not* prove that the inverse function is differentiable, which is what the above theorem proved. In fact, in using chain rule, we are already assuming that g is differentiable.

Example 5.4. For a positive integer n, the function $f(x) = x^n$ is strictly increasing and differentiable on $\mathbb{R}$ if n is odd and on $(0, \infty)$ if n is even. Its derivative is $f'(x) = nx^{n-1}$. Its inverse function is the nth root function $g(x) = x^{\frac{1}{n}}$, which is only defined for $x > 0$ if n is even. Writing $0 \neq y = f(x) = x^n$ we have $g'(y) = \frac{1}{nx^{n-1}}$, which yields

$$g'(x) = \frac{1}{nx^{\frac{n-1}{n}}} = \frac{1}{n} x^{\frac{1-n}{n}} = \frac{1}{n} x^{\frac{1}{n} - 1}.$$

Therefore, the derivative formula for x^n, where n is a positive integer, still holds if n is replaced by $\frac{1}{n}$.

5.1.6   *Exercises*

(1) Use the definition of derivative to find the derivatives, when they exist, of the following functions.
    (a) $\sqrt{2x + 5}$
    (b) $\frac{1}{\sqrt{x+4}}$
    (c) $2x^3 + 10x + 5$
    (d) $\frac{1}{x^2+1}$

(e) $\frac{x}{x^2+1}$

(2) Write down the sequential and the ϵ-δ characterizations of the derivative. Write down what it means for f to not be differentiable at c using ϵ-δ.

(3) Prove that a constant function is differentiable on $\mathbb{R}$, and compute its derivative 0.

(4) Prove the first three parts of Theorem 5.2.

(5) Prove that a polynomial is differentiable on $\mathbb{R}$.

(6) Suppose $r = \frac{p}{q}$ is a rational number in lowest terms and $g(x) = x^r$. Prove that g is differentiable on $\mathbb{R} \setminus \{0\}$ if q is odd and on $(0, \infty)$ if q is even. Then show that $g'(x) = rx^{r-1}$.

(7) Suppose f and g are both differentiable at c and that $f(x) \le g(x)$ for all x in some open interval containing c. Does it follow that $f'(c) \le g'(c)$?

(8) Prove the **generalized product rule**

$$(fg)^{(n)}(c) = \sum_{k=0}^{n} \binom{n}{k} f^{(k)}(c) g^{(n-k)}(c),$$

assuming all the derivatives exist, where $\binom{n}{k}$ is the binomial coefficient in (1.4).

(9) Define the function

$$f(x) = \begin{cases} x \sin\left(\frac{1}{x}\right) & \text{if } x \neq 0, \\ 0 & \text{if } x = 0. \end{cases}$$

Prove that f is not differentiable at 0.

(10) Define the function

$$f(x) = \begin{cases} x^2 \sin\left(\frac{1}{x}\right) & \text{if } x \neq 0, \\ 0 & \text{if } x = 0. \end{cases}$$

Prove that f is differentiable at 0, and compute $f'(0)$. You may assume that the function $\sin(x)$ is differentiable on $\mathbb{R}$ and that its derivative is $\cos(x)$. Show that f' is not continuous at 0.

(11) Prove that $f(x) = |x|$ is differentiable on $(-\infty, 0)$ and $(0, \infty)$, and compute its derivative.

(12) If f is differentiable at c, prove that

$$f'(c) = \lim_{h \to 0} \frac{f(c+h) - f(c)}{h} = \lim_{h \to 0} \frac{f(c+h) - f(c-h)}{2h}.$$

Draw pictures to visualize these limits in terms of secant lines.

(13) In the previous exercise, if the last limit exists, does it follow that f is differentiable at c?

(14) Let $f \colon \mathbb{R} \to \mathbb{R}$ be the function $f(x) = x^2$ if x is rational and $f(x) = 0$ if x is irrational. Prove that f is differentiable at 0, and compute $f'(0)$.

(15) Give an example of a differentiable function f on an open interval I that is not uniformly continuous.

(16) Suppose f is differentiable on an open interval I and that f' is bounded on I. Prove that f is uniformly continuous on I.

5.2   Mean Value Theorem

The title of this section refers to an important theorem that says that the slope of the secant line of a function on a closed interval is, in fact, the slope of some tangent line.

5.2.1   *Two Preliminary Theorems*

In order to prove the Mean Value Theorem, we need two preliminary observations. The following observation says that if a function achieves an extremum at a point where it is differentiable, then its derivative must be zero there. Intuitively, this is true because, if the derivative is not zero, then the function is either increasing or decreasing there. But then that point cannot be an extremum. The proof is a formal version of this line of reasoning.

Theorem 5.5 (Interior Extremum Theorem). *Suppose f is defined on an open interval I, c is a point in I where f is differentiable, and f achieves its maximum or minimum on I at c. Then $f'(c) = 0$.*

Ideas of Proof. If $f'(c) \neq 0$, then there are points close to c where f is larger or smaller than $f(c)$, which would contradict the maximum or minimum assumption.

Proof.   Say f achieves its maximum at c. The minimum case is left as an exercise. Suppose to the contrary that $f'(c) \neq 0$, say, $f'(c) > 0$. Then there exists $\delta > 0$ such that

$$0 < |x - c| < \delta \text{ with } x \in I \quad \text{implies} \quad \frac{f(x) - f(c)}{x - c} > 0.$$

Choosing any $x > c$ in I, the above inequality then implies

$$f(x) > f(c),$$

which contradicts the maximum assumption. If $f'(c) < 0$, then likewise there exists $\delta > 0$ such that

$$0 < |x - c| < \delta \text{ with } x \in I \quad \text{implies} \quad \frac{f(x) - f(c)}{x - c} < 0.$$

Choosing any $x < c$ in I, the above inequality implies

$$f(x) > f(c),$$

which is again a contradiction.   □

The following observation is a special case of the Mean Value Theorem. It says roughly that if a function has equal values at two distinct points, then its derivative must be zero somewhere between those two points. Intuitively, this is easy to see if one tries to draw a graph of such a function.

Theorem 5.6 (Rolle's Theorem). *Suppose f is continuous on a closed interval $[a, b]$, is differentiable on (a, b), and $f(a) = f(b)$. Then there exists a point*

$$c \in (a, b) \quad \text{such that} \quad f'(c) = 0.$$

Proof. By the Interior Extremum Theorem, if f achieves either a maximum or a minimum on (a, b) at a point c, then $f'(c) = 0$. By the Extreme Value Theorem 4.6, f achieves both a maximum and a minimum on $[a, b]$. If either one of them is achieved on (a, b), then we are done by the Interior Extremum Theorem. Otherwise, f achieves its maximum and minimum at a or b. But then the assumption $f(a) = f(b)$ implies that f is constant on $[a, b]$. Therefore, $f'(c) = 0$ for every point c in (a, b). □

5.2.2 Main Result

We are now ready for the main result of this section.

Theorem 5.7 (Mean Value Theorem). *Suppose f is continuous on a closed interval $[a, b]$ and differentiable on (a, b). Then there exists a point c in (a, b) such that*

$$f'(c) = \frac{f(b) - f(a)}{b - a}.$$

Ideas of Proof. The plan is to use Rolle's Theorem by subtracting from f the secant line connecting the points at $x = a$ and $x = b$, whose equation is

$$y = f(b) + \frac{f(b) - f(a)}{b - a}(x - b).$$

Proof. Define the function

$$g(x) = f(x) - f(b) - \frac{f(b) - f(a)}{b - a}(x - b),$$

which is continuous on $[a, b]$ and differentiable on (a, b). Since

$$g(a) = g(b) = 0,$$

Rolle's Theorem implies that there exists a point c in (a, b) such that

$$0 = g'(c) = f'(c) - \frac{f(b) - f(a)}{b - a},$$

as desired. □

The reader should draw a typical graph and try to interpret the Mean Value Theorem using the graph. In any case, the quotient on the right-hand side is the slope of a secant line, and $f'(c)$ is the slope of a tangent line somewhere in between.

Note that Rolle's Theorem is a special case of the Mean Value Theorem because the quotient in the latter is 0 when $f(a) = f(b)$.

The Mean Value Theorem, and its special case Rolle's Theorem, are ultimately about showing that certain quantity is in the range of the derivative. One way of using them is to establish certain inequalities that are otherwise hard to prove, as the following example illustrates.

Example 5.5. Here we show that

$$ex \leq e^x$$

for all x in $\mathbb{R}$. If $f(x) = e^x - ex$, then we must show $f(x) \geq 0$. Note that the only root of $f(x) = 0$ is $x = 1$. Indeed, if there is another root $r \neq 1$, then Rolle's Theorem says $f'(d) = 0$ for some point d in between r and 1. But the only root of

$$f'(x) = e^x - e = 0$$

is $x = 1$. Moreover, $f'(x) < 0$ if $x < 1$ and $f'(x) > 0$ if $x > 1$, so f is strictly decreasing on $(-\infty, 1)$ and strictly increasing on $(1, \infty)$. Since $x = 1$ is the only zero of f, we conclude that $f(x) \geq 0$ for all x.

5.2.3 Consequences

We now discuss some consequences of the Mean Value Theorem. The following observation says that a function whose derivative is identically 0 is a constant function. Intuitively, having derivative 0 implies that the function can neither increase nor decrease, which means it is constant.

Corollary 5.1. *Suppose f is differentiable on (a, b) with $f' = 0$ on (a, b). Then f is a constant function.*

Proof. If f is not a constant function, then there exist $c < d$ in (a, b) such that $f(c) \neq f(d)$. Applying the Mean Value Theorem to f on $[c, d]$, we obtain a point e in (c, d) such that

$$f'(e) = \frac{f(d) - f(c)}{d - c} \neq 0,$$

which contradicts the assumption $f' = 0$. □

The next observation says that if two functions have the same derivative on some interval, then they can differ by at most a constant. Intuitively, these two functions have to increase or decrease at exactly the same rate everywhere on this interval. So they differ by the same amount at all the points on this interval.

Corollary 5.2. *Suppose f and g are both differentiable on (a, b) with $f' = g'$ on (a, b). Then there exists a constant C such that $f = g + C$ on (a, b).*

Proof.   The difference $h = f - g$ is differentiable on (a, b) and satisfies $h' = 0$, so the previous corollary says h is a constant function. In other words, $f = g + C$ for some constant C.   □

The previous result is important in integration. It says that if a function f has an anti-derivative F on an interval, then $\{F + C : C \text{ constant}\}$ is the entire family of anti-derivatives of f. Therefore, to find all the anti-derivatives of a function, it suffices to find just one of them and form the above family.

More consequences of the Mean Value Theorem are explored in the exercises and the next section.

5.2.4   *Exercises*

(1) Finish the proof of Theorem 5.5 by proving the case when f achieves its minimum at c.

(2) Give a proof of Bernoulli's Inequality (Theorem 1.5) using the Mean Value Theorem.

(3) Prove the inequality $|\cos x - \cos y| \le |x - y|$ for all real numbers x and y.

(4) Prove the inequality $|\sin x - \sin y| \le |x - y|$ for all real numbers x and y.

(5) Suppose f is differentiable on an open interval I. Prove the following statements.

    (a) If $f'(x) > 0$ for all x in I, then f is strictly increasing.

    (b) If $f'(x) \ge 0$ for all x in I, then f is increasing.

    (c) If $f'(x) < 0$ for all x in I, then f is strictly decreasing.

    (d) If $f'(x) \le 0$ for all x in I, then f is decreasing.

(6) Suppose f and g are both differentiable, and $f(a) = g(a)$.

    (a) If $f'(x) \le g'(x)$ for all $x \ge a$, prove that $f(x) \le g(x)$ for all $x \ge a$.

    (b) If $f'(x) < g'(x)$ for all $x > a$, prove that $f(x) < g(x)$ for all $x > a$.

(7) Prove that $\ln(1 + x) < x$ for all $x > 0$.

(8) Prove that $|\sin x| \le x$ for all $x \ge 0$.

(9) Prove that $\tan x > x$ if $0 < x < \frac{\pi}{2}$.

(10) Prove that $e^x > 1 + x$ for all $x > 0$.

(11) Suppose f is differentiable on an open interval I with f' bounded. Prove that there exists a real number $M > 0$ such that

$$|f(x) - f(y)| \le M|x - y|$$

for all x, y in I.

(12) Suppose f is continuous on $[a, b]$, differentiable on (a, b), and $f(c) > f(a) = f(b)$ for some point c in (a, b). Prove that there exist two points x and y in (a, b) such that $f'(x) < 0 < f'(y)$.

(13) Suppose f is continuous on $[0, 1]$, differentiable on $(0, 1)$, $f(0) = 0$, and f' is increasing on $(0, 1)$. Prove that $g(x) = \frac{f(x)}{x}$ is increasing on $(0, 1)$.

(14) Suppose $0 < x < y$ and $n \geq 2$ is an integer. Prove that $y^{\frac{1}{n}} - x^{\frac{1}{n}} < (y - x)^{\frac{1}{n}}$.

(15) Suppose $x, y > 0$ and $0 < r < 1$. Prove that $x^r y^{1-r} \leq rx + (1-r)y$.

(16) Suppose f is continuously differentiable on an open interval I and a is a point in I such that $f(a) = 0$ and $f'(a) \neq 0$. Prove that there exists an open interval $J \subseteq I$ containing a on which both f and f' are non-zero with the exception of $f(a) = 0$.

5.3   Taylor's Theorem

In this section we discuss other important results related to the Mean Value Theorem, the main one being Taylor's Theorem. We then discuss L'Hospital's Rule and the intermediate value property for derivatives.

5.3.1   *Motivation*

Let us discuss the motivation behind the statement of Taylor's Theorem. Polynomials are among the easiest functions to work with. For example, they have very simple derivative formulas, and the product of two polynomials remains a polynomial. Moreover, computing a polynomial involves only the arithmetic operations of addition, subtraction, and multiplication, which can be easily handled by computers. Therefore, it makes sense to approximate other functions using polynomials. Moreover, if we want to understand a function f near a point c, then the polynomials should be in powers of $(x - c)$ instead of x.

Suppose it is possible to write

$$f(x) = a_0 + a_1(x - c) + a_2(x - c)^2 + a_3(x - c)^3 + \cdots$$

in powers of $(x - c)$. Then

$$f(c) = a_0, \quad f'(c) = a_1, \quad f''(c) = 2a_2, \quad f'''(c) = 3!a_3,$$

and so forth. The general formula for the kth coefficient is

$$a_k = \frac{f^{(k)}(c)}{k!}.$$

Of course, f does not need to be a polynomial, so the above expression of f should have an error term. In other words, a degree n polynomial approximation of f should take the form

$$f(x) = T_n(x; c) + R_n(x; c), \tag{5.4}$$

where

$$T_n(x; c) = f(c) + f'(c)(x - c) + \cdots + \frac{f^{(n)}(c)}{n!}(x - c)^n$$

$$= \sum_{k=0}^{n} \frac{f^{(k)}(c)}{k!}(x - c)^k \tag{5.5}$$

is a degree n polynomial in $(x - c)$, and $R_n(x; c)$ is the nth **error term**. The polynomial $T_n(x; c)$ is called the **degree n Taylor polynomial** of f at c. Taylor's Theorem provides an actual expression for the error term in terms of a higher derivative.

5.3.2 *Main Result*

We now state and prove **Taylor's Theorem**.

Theorem 5.8. *Suppose n is a positive integer and $f:[\alpha, \beta] \to \mathbb{R}$ is a function such that $f^{(k)}$ is continuous on $[\alpha, \beta]$ for $0 \le k \le n$ and that $f^{(n+1)}$ exists on (α, β). Then given two distinct points c and x in $[\alpha, \beta]$, there exists a point b strictly between them such that*

$$f(x) = T_n(x; c) + \frac{f^{(n+1)}(b)}{(n+1)!}(x - c)^{n+1}.$$

Ideas of Proof. We are trying to show that certain quantity is in the range of the higher derivative $f^{(n+1)}$. Therefore, as in the proof of the Mean Value Theorem, we are going to use Rolle's Theorem on a suitable function.

Proof. We need to show that, in the equality

$$\frac{K}{(n+1)!}(x - c)^{n+1} = f(x) - T_n(x; c),$$

the quantity K must be $f^{(n+1)}(b)$ for some b strictly between c and x. To prove this, consider the function in t

$$F(t) = f(x) - T_n(x; t) - \frac{K}{(n+1)!}(x - t)^{n+1}$$

defined on the closed interval J formed by c and x. The function $F(t)$ is continuous on J, differentiable between c and x, and satisfies $F(c) = 0 = F(x)$ by the definition of K and by $T_n(x; x) = f(x)$. Therefore, Rolle's Theorem 5.6 yields a point b strictly between c and x such that $F'(b) = 0$. The reader can check with a direct computation that

$$\frac{d}{dt}(T_n(x; t)) = \frac{f^{(n+1)}(t)}{n!}(x - t)^n. \tag{5.6}$$

Therefore, we have

$$0 = F'(b) = -\frac{f^{(n+1)}(b)}{n!}(x - b)^n - \frac{K(n+1)}{(n+1)!}(x - b)^n(-1),$$

which simplifies to $K = f^{(n+1)}(b)$, as desired. $\qquad\square$

In terms of the error term in (5.4), Taylor's Theorem says

$$R_n(x;c) = \frac{f^{(n+1)}(b)}{(n+1)!}(x-c)^{n+1} \tag{5.7}$$

for some point b between c and x. In other words, the error term is given in terms of the $(n+1)$st derivative of f. Another form of the error term is discussed in Theorem 6.10.

One typical use of Taylor's Theorem is to approximate functions using low degree polynomials, as in the next example.

Example 5.6. Suppose $f(x) = \sin x$ and $a = 0$. For any $x \neq 0$, Taylor's Theorem with $n = 3$ yields

$$\sin x = T_3(x;0) + \frac{f^{(4)}(c)}{4!}x^4 = x - \frac{x^3}{3!} + \frac{\sin c}{4!}x^4$$

for some c strictly between 0 and x. Since $|\sin x| \leq 1$ for all x, the error term is bounded above by $\frac{x^4}{4!}$, which is a very small number for x near 0. For example, if $x = 0.1$, then $\frac{(0.1)^4}{4!}$ is about 4.2×10^{-6}.

5.3.3   L'Hospital's Rule

Next we discuss a simple version of L'Hospital's Rule, which the reader must have encountered in differential calculus. The result says roughly that, if $f(a) = 0 = g(a)$, then $\lim_{x \to a} \frac{f}{g}$ is equal to $\frac{f'(a)}{g'(a)}$ if it exists.

Theorem 5.9. *Suppose f and g are both continuously differentiable on an open interval I, a is a point in I such that $f(a) = 0 = g(a)$, and that $g'(a) \neq 0$. Then there is an equality*

$$\lim_{x \to a} \frac{f(x)}{g(x)} = \frac{f'(a)}{g'(a)}.$$

Ideas of Proof. Near the point a, the graphs of f and g are approximated by their respective tangent lines. Since $f(a) = 0 = g(a)$, these lines have equations

$$y = f'(a)(x-a) \quad \text{and} \quad y = g'(a)(x-a),$$

respectively, which are also the degree 1 Taylor polynomials of f and g. Therefore, near the point a, the quotient $\frac{f}{g}$ is roughly equal to

$$\frac{f'(a)(x-a)}{g'(a)(x-a)} = \frac{f'(a)}{g'(a)}.$$

Proof. There is an open interval $J \subseteq I$ containing a on which g and g' are non-zero with the exception of $g(a) = 0$ (Exercise (16) on page 126). For $x \neq a$ in J, the Mean Value Theorem says

$$\frac{f(x)}{g(x)} = \frac{\left(\frac{f(x)-f(a)}{x-a}\right)}{\left(\frac{g(x)-g(a)}{x-a}\right)} = \frac{f'(c)}{g'(d)}$$

for some c and d strictly between a and x. To finish the proof, just take the limit $\lim_{x \to a}$, using the assumption that f' and g' are continuous. □

Example 5.7. We have

$$\lim_{x \to 1} \frac{\ln x}{x - 1} = 1$$

by L'Hospital's Rule, since $(\ln x)' = \frac{1}{x}$ and $(x - 1)' = 1$.

There is a more general form of L'Hospital's Rule involving higher derivatives. It will be explored in the exercises.

5.3.4 *Intermediate Value Theorem for Derivative*

Recall that a continuous function on a closed bounded interval satisfies the Intermediate Value Theorem 4.7. On the other hand, a differentiable function does not need to have a continuous derivative, so the Intermediate Value Theorem does not apply to the derivative in general. Nevertheless, the next observation says that the derivative does satisfy the conclusion of the Intermediate Value Theorem. Its proof uses the Interior Extremum Theorem instead of the Mean Value Theorem.

Theorem 5.10. *Suppose f is differentiable on (a, b), and r is a number that lies strictly between $f'(c)$ and $f'(d)$ for some $c < d$ in (a, b). Then there exists a point*

$$e \in (c, d) \quad \text{such that} \quad r = f'(e).$$

Ideas of Proof. The plan is similar to the proof of the Mean Value Theorem. We will subtract from f the line through the origin with slope r and apply the Interior Extremum Theorem.

Proof. Let us consider the case $f'(c) < r < f'(d)$. The other case is very similar. Consider the function

$$g(x) = f(x) - rx,$$

which is differentiable on (a, b) with

$$g'(x) = f'(x) - r.$$

In particular, it suffices to show that there exists a point e in (c, d) with $g'(e) = 0$. Thus, by the Interior Extremum Theorem, it is enough to show that g achieves either a maximum or a minimum on (c, d) at some point e. By the Extreme Value Theorem 4.6, g achieves a minimum on $[c, d]$ at some point e. To apply the Interior Extremum Theorem, we just need to observe that e is neither c nor d. In other words, we need to show that

$$g(x_1) < g(c) \quad \text{and} \quad g(x_2) < g(d)$$

for some x_1 and x_2 in (c, d). Note that

$$g'(c) < 0 < g'(d).$$

So there exists $\delta_1 > 0$ such that

$$0 < |x - c| < \delta_1 \text{ with } x \in (a, b) \quad \text{implies} \quad \frac{g(x) - g(c)}{x - c} < 0,$$

which in turn implies

$$g(x) < g(c) \quad \text{if} \quad x > c.$$

Therefore, e is not the end point c. Likewise, the inequality $0 < g'(d)$ implies that there exists $\delta_2 > 0$ such that

$$0 < |x - d| < \delta_2 \text{ with } x \in (a, b) \quad \text{implies} \quad \frac{g(x) - g(d)}{x - d} > 0.$$

This last inequality implies

$$g(x) < g(d) \quad \text{if} \quad x < d,$$

so e is not the end point d. $\qquad\qquad\qquad\qquad\qquad\qquad\qquad\qquad\qquad\square$

5.3.5   Exercises

(1) Prove the equality (5.6).

(2) Finish the proof of Theorem 5.10 by proving the case when $f'(d) < r < f'(c)$.

(3) Use Taylor's Theorem to approximate $\sqrt{5}$ to within 0.0001.

(4) Use Taylor's Theorem to approximate $\cos(0.1)$ to within 0.0001.

(5) Use Taylor's Theorem to approximate $\ln 2$ to within 0.001.

(6) Prove the inequalities

$$1 - \frac{x^2}{2} \le \cos x \le 1 - \frac{x^2}{2} + \frac{x^4}{4!}$$

for $x > 0$.

(7) Prove the inequalities

$$x - \frac{x^3}{3!} \le \sin x \le x - \frac{x^3}{3!} + \frac{x^5}{5!}$$

for $x > 0$.

(8) Prove the inequalities

$$\sum_{k=1}^{2N} (-1)^{k-1} \frac{x^k}{k} < \ln(1+x) < \sum_{k=1}^{2N+1} (-1)^{k-1} \frac{x^k}{k}$$

for each positive integer N and each positive real number x.

(9) Prove the inequalities

$$\sqrt{2} + \frac{x}{2\sqrt{2}} - \frac{x^2}{8\sqrt{2}} < \sqrt{2+x} < \sqrt{2} + \frac{x}{2\sqrt{2}}$$

for $x > 0$.

(10) Prove the inequality

$$e^x > 1 + x + \frac{x^2}{2!} + \cdots + \frac{x^n}{n!}$$

for $x > 0$ and $n \geq 1$.

(11) Compute the limits.

(a) $\lim_{x \to 0} \frac{\sin 4x}{x}$

(b) $\lim_{x \to 1} \frac{x^9 - 1}{x^7 - 1}$

(c) $\lim_{x \to 1} \frac{e^x - e}{e^{2x} - e^2}$

(d) $\lim_{x \to 1} \frac{\ln x}{\cos(\frac{\pi x}{2})}$

(12) Prove the following generalization of Exercise (16) on p. 126. Suppose $n \geq 1$, f has continuous $(n + 1)$st derivative on an open interval I, and a is a point in I such that

$$0 = f(a) = f'(a) = \cdots = f^{(n)}(a) \quad \text{and} \quad f^{(n+1)}(a) \neq 0.$$

Then there exists an open interval $J \subseteq I$ containing a on which $f^{(k)}$ is non-zero for $0 \leq k \leq n + 1$ with the exception of the point a.

(13) Prove the following generalization of L'Hospital's Rule. Suppose $n \geq 1$, f and g both have continuous $(n+1)$st derivatives on an open interval I, a is a point in I such that $0 = f^{(k)}(a) = g^{(k)}(a)$ for $0 \leq k \leq n$, and that $g^{(n+1)}(a) \neq 0$. Then

$$\lim_{x \to a} \frac{f(x)}{g(x)} = \frac{f^{(n+1)}(a)}{g^{(n+1)}(a)}.$$

(14) Compute the limits

(a) $\lim_{x \to 0} \frac{\cos x - 1}{x^2}$

(b) $\lim_{x \to 0} \frac{\tan^2 x}{x^2}$

(c) $\lim_{x \to 0} \frac{\tan x - x}{x^3}$

5.4 Additional Exercises

(1) Give an example of a differentiable, uniformly continuous function f on some open interval such that f' is unbounded.

(2) Formulate and prove a version of chain rule involving the composition of three functions.

(3) Prove that $f(x) = |x^3|$ is differentiable on $(-\infty, 0)$ and $(0, \infty)$, and compute its derivative.

(4) Let n be a positive integer. Suppose $f: \mathbb{R} \to \mathbb{R}$ is the function $f(x) = x^n$ if x is rational and $f(x) = 0$ if x is irrational. Prove that f is differentiable at 0, and compute $f'(0)$.

(5) Suppose $f''(c)$ exists. Prove that

$$f''(c) = \lim_{h \to 0} \frac{f(c+h) - 2f(c) + f(c-h)}{h^2}.$$

Give an example of a function for which this limit exists, but $f''(c)$ does not exist.

(6) A function $f:\mathbb{R} \to \mathbb{R}$ is called **odd** if $f(-x) = -f(x)$ for all x and is called **even** if $f(-x) = f(x)$ for all x. Prove the following statements.

(a) If f is a differentiable even function, then f' is odd.
(b) If f is a differentiable odd function, then f' is even.

(7) Prove **Carathéodory's Theorem**: Suppose f is defined on an open interval I containing c. Then f is differentiable at c if and only if there exists a function g defined on I that is continuous at c and satisfies

$$f(x) - f(c) = g(x)(x - c)$$

for $x \in I$. When f is differentiable at c, prove that $g(c) = f'(c)$.

(8) The **left-hand derivative** of f at c is defined as the left-hand limit

$$\lim_{x \to c^-} \frac{f(x) - f(c)}{x - c},$$

if it exists. The **right-hand derivative** is defined similarly using the right-hand limit. Prove that f is differentiable at c if and only if its left-hand and right-hand derivatives both exist and are equal.

(9) Suppose f is continuous on $[a,b]$, differentiable on (a,b), $f(a) = f(b) = 0$, and r is a real number. Prove that there exists a point c in (a,b) such that $f'(c) = rf(c)$.

(10) **Cauchy's Mean Value Theorem** says: Suppose f and g are both continuous on $[a,b]$ and differentiable on (a,b). Then there exists a point c in (a,b) such that

$$(f(b) - f(a))g'(c) = (g(b) - g(a))f'(c).$$

Prove this theorem as follows.

(a) First consider the case $g(a) = g(b)$, and use Rolle's Theorem.
(b) Next consider the case $g(a) \neq g(b)$. The required equality is equivalent to

$$\frac{f'(c)}{g'(c)} = \frac{f(b) - f(a)}{g(b) - g(a)}.$$

Let r denote the quotient on the right-hand side, and define $h = f - rg$. Now show that Rolle's Theorem applies to h.

(11) Give another proof of the Mean Value Theorem using Cauchy's Mean Value Theorem.

(12) Suppose f is defined on an open interval I such that

$$|f(x) - f(y)| \leq M(x - y)^2$$

for some $M > 0$ and all x, y in I.

(a) Prove that f is uniformly continuous on I.
(b) Prove that f is differentiable on I.

(c) Prove that f is a constant function on I.

(13) Suppose f is differentiable on an open interval such that f' is strictly monotone. Prove that f' is continuous.

(14) Suppose

$$f(x) = \begin{cases} x^4 \left(2 + \sin\left(\frac{1}{x}\right)\right) & \text{if } x \neq 0, \\ 0 & \text{if } x = 0. \end{cases}$$

Prove the following statements.

(a) f achieves an absolute minimum at 0.

(b) In each open interval containing 0, f' takes on both positive and negative values.

This example shows that the converse of the first derivative test in differential calculus is false.

(15) Suppose f'' exists on (a, b), and $x < y < z$ are three points in (a, b) such that both $f(x)$ and $f(z)$ are strictly greater than $f(y)$. Prove that there exists a point c in (a, b) such that $f''(c) > 0$.

Chapter 6

Integration

In this chapter we study integration, which is in some sense a reverse process of differentiation. Integrals are defined in section 6.1 using lower and upper sums. An alternative approach to integrals based on tagged partitions is given in section 6.2. Basic properties of integrals, including the Mean Value Theorem for integrals, are given 6.3. The Fundamental Theorem of Calculus and some of its important consequences are proved in section 6.4.

6.1 The Integral

In this section, we first discuss lower and upper sums associated to partitions. The integral is defined in terms of these lower and upper sums. We then prove a useful ϵ-criterion for integrability and use it to show that continuous functions are integrable.

6.1.1 *Motivation*

Let us briefly discuss the motivation behind the definitions below. As one learns in calculus, the definite integral is intuitively the area under the graph of a function over an interval. To define it algebraically, one uses a similar process as for derivative, which in nice cases represents the slope of the tangent. More precisely, in the definition of the derivative, the difference quotient $\frac{f(x)-f(c)}{x-c}$ represents the slope of a secant line that is close to the tangent at c. Taking the limit as $x \to c$ then yields the slope of the tangent at c for nice functions.

The integral for nice functions can be intuitively represented as the area under the graph. To approximate it, we first cut the interval into several sub-intervals. Over each small sub-interval, we can approximate the area using a rectangle whose height is the value of the function at a chosen point. One has choices as to which points to use for the heights. Two obvious choices are the maximum and the minimum, which lead to circumscribed and inscribed rectangles. We can then define the area/integral in terms of these rectangles. Since these approximations yield sets

135

of numbers rather than functions, we will replace $\lim_{x \to c}$ in the derivative with suitable infimum and supremum. The definitions below are formal versions of these ideas.

6.1.2   *Upper and Lower Sums*

Throughout the rest of this chapter, unless otherwise specified, f is a bounded function defined on a closed interval $[a, b]$. Recall from Definition 4.9 the concept of a partition of $[a, b]$.

Definition 6.1. Suppose $P = \{a = x_0 < x_1 < \cdots < x_n = b\}$ is a partition of $[a, b]$.

(1) Its **norm** is defined as

$$\|P\| = \max\{x_i - x_{i-1} : i = 1, \ldots, n\}.$$

(2) Define

- $\Delta x_i = x_i - x_{i-1}$,
- $l_i(f) = \inf\{f(x) : x \in [x_{i-1}, x_i]\}$, and
- $u_i(f) = \sup\{f(x) : x \in [x_{i-1}, x_i]\}$

  for $i = 1, \ldots, n$.

(3) The **lower sum of f with respect to P** is the sum

$$L(f; P) = \sum_{i=1}^{n} l_i(f) \Delta x_i.$$

(4) The **upper sum of f with respect to P** is the sum

$$U(f; P) = \sum_{i=1}^{n} u_i(f) \Delta x_i.$$

(5) If Q is another partition of $[a, b]$ such that $P \subseteq Q$, then Q is called a **refinement** of P.

   The reader should be careful that the infimum $l_i(f)$ and the supremum $u_i(f)$ may not be in the range of f. So the lower sum and the upper sum with respect to a partition are not necessarily Riemann sums in the calculus sense.

Definition 6.2. The **lower sum** and the **upper sum** of f on $[a, b]$ are defined as the least upper bound

$$L(f) = \sup\{L(f; P) : P \text{ a partition of } [a, b]\}$$

and the greatest lower bound

$$U(f) = \inf\{U(f; P) : P \text{ a partition of } [a, b]\},$$

respectively.

A major confusing point about the lower sum and the upper sum is that each of them involves both infimum and supremum. For example, to define the lower sum, first we form the lower sum with respect to a partition P by using the infimum $l_i(f)$ on each sub-interval of P. Then we take the supremum of the lower sums with respect to the partitions. Likewise, the upper sum involves the supremum $u_i(f)$ on each sub-interval of a partition and then the infimum of the upper sums with respect to partitions. Notice that if we switch the order of the infimum and the supremum in the definition of the lower sum, the result is the upper sum, and vice versa. As we will see shortly below, being integrable is essentially saying that the infimum and the supremum operations commute.

To define the integral, we first need to establish some basic properties of the lower sum and the upper sum.

Theorem 6.1. *Suppose f is a bounded function on $[a,b]$, and $P \subseteq Q$ are partitions of $[a,b]$. Then*

$$m(b-a) \le L(f;P) \le L(f;Q) \le U(f;Q) \le U(f;P) \le M(b-a),$$

where $m = \inf\{f(x) : x \in [a,b]\}$ and $M = \sup\{f(x) : x \in [a,b]\}$.

Proof. For the left-most inequality, we have $m \le l_i(f)$ for each i, so

$$m(b-a) = \sum_{i=1}^{n} m\Delta x_i \le \sum_{i=1}^{n} l_i(f)\Delta x_i = L(f;P).$$

For the second inequality, first note that Q has only finitely many points more than P. Thus, by an induction argument it is enough to prove the case where Q has exactly one partition point $t \in (x_{i-1}, x_i)$ more than P, where x_{i-1} and x_i are two consecutive partition points in P. Denote the infimum of f on $[x_{i-1}, t]$ and $[t, x_i]$ by $l_i'(f)$ and $l_i''(f)$, respectively. Since

$$\Delta x_i = (x_i - t) + (t - x_{i-1}),$$

we have

$$l_i(f)\Delta x_i = l_i(f)(t - x_{i-1}) + l_i(f)(x_i - t)$$
$$\le l_i'(f)(t - x_{i-1}) + l_i''(f)(x_i - t)$$

because $l_i(f) \le l_i'(f)$ and $l_i(f) \le l_i''(f)$. Now on $[a, x_{i-1}]$ and $[x_i, b]$, the partitions P and Q have the same partition points, so the previous inequality implies the desired inequality

$$L(f;P) \le L(f;Q).$$

The third inequality holds because $l_i(f) \le u_i(f)$ for each i. The other two inequalities are left as exercises for the reader. $\square$

The previous result actually implies that each lower sum is no more than each upper sum with respect to arbitrary partitions. Intuitively, a lower sum with respect to some partition consists of inscribed rectangles, so the sum of their areas cannot exceed that of circumscribed rectangles with respect to any other partition.

Corollary 6.1. *Suppose f is a bounded function on $[a,b]$, and P and Q are two partitions of $[a,b]$. Then*

$$L(f;P) \le U(f;Q).$$

Proof.   The union $P \cup Q$ is a refinement of both P and Q, so the previous theorem yields

$$L(f;P) \le L(f;P \cup Q) \le U(f;P \cup Q) \le U(f;Q), \tag{6.1}$$

as desired.   □

Corollary 6.2. *Suppose f is a bounded function on $[a,b]$, and P and Q are two partitions of $[a,b]$. Then*

$$L(f;P) \le L(f) \le U(f) \le U(f;Q).$$

Proof.   The two outer inequalities hold by the definitions of the lower sum and the upper sum. For the middle inequality, we know that $L(f;P)$ is a lower bound of every $U(f;Q)$ by Corollary 6.1. Since $U(f)$ is the greatest lower bound of these $U(f;Q)$, we have

$$L(f;P) \le U(f).$$

This inequality says that $U(f)$ is an upper bound of $L(f;P)$. Since P is arbitrary, the middle inequality follows.   □

6.1.3   *Integrability*

Definition 6.3. A bounded function f on $[a,b]$ is said to be **integrable** on $[a,b]$ if its lower sum and upper sum are equal, that is,

$$L(f) = U(f).$$

In this case, the common value is denoted by

$$\int_a^b f \quad \text{or} \quad \int_a^b f(x)dx,$$

and it is called the **integral** of f on $[a,b]$. If f is integrable on $[a,b]$, we define

$$\int_b^a f = -\int_a^b f.$$

From now on, when we say f is integrable on $[a,b]$, we automatically assume that f is bounded on $[a,b]$. When f is integrable on $[a,b]$, the dummy variable x in $\int_a^b f(x)dx$ may be changed to any other symbol, such as u. We will discuss several classes of integrable functions. Before that, let us first exhibit a function that is not integrable.

Example 6.1. Here we observe that the characteristic function $\chi_{\mathbb{Q}}$ of $\mathbb{Q}$ defined in (4.2) is not integrable on any interval $[a, b]$. Indeed, if P is a partition of $[a, b]$, then in each sub-interval $[x_{i-1}, x_i]$ there are both rational and irrational numbers. So

$$l_i(\chi_{\mathbb{Q}}) = 0 \quad \text{and} \quad u_i(\chi_{\mathbb{Q}}) = 1$$

for each i. This implies

$$L(\chi_{\mathbb{Q}}; P) = 0 \quad \text{and} \quad U(\chi_{\mathbb{Q}}; P) = \sum \Delta x_i = b - a.$$

Since P is an arbitrary partition, we have

$$L(\chi_{\mathbb{Q}}) = 0 \quad \text{and} \quad U(\chi_{\mathbb{Q}}) = b - a,$$

which shows that $\chi_{\mathbb{Q}}$ is not integrable on $[a, b]$.

6.1.4 *Criterion for Integrability*

To establish integrability of functions, the following ϵ-criterion is often useful.

Theorem 6.2. *A bounded function f on $[a, b]$ is integrable if and only if, for each $\epsilon > 0$, there exists a partition P of $[a, b]$ such that*

$$U(f; P) - L(f; P) < \epsilon.$$

Proof. The "only if" part is an $\frac{\epsilon}{2}$-argument. Suppose f is integrable on $[a, b]$, so $I = \int_a^b f = L(f) = U(f)$. Given $\epsilon > 0$, we have

$$I - \frac{\epsilon}{2} < L(f) \quad \text{and} \quad I + \frac{\epsilon}{2} > U(f),$$

so there exist partitions P_1 and P_2 such that

$$L(f; P_1) > I - \frac{\epsilon}{2} \quad \text{and} \quad U(f; P_2) < I + \frac{\epsilon}{2}. \tag{6.2}$$

If $P = P_1 \cup P_2$, then (6.1) and (6.2) imply

$$U(f; P) - L(f; P) \le U(f; P_2) - L(f; P_1) < \frac{\epsilon}{2} + \frac{\epsilon}{2} = \epsilon,$$

proving the "only if" part.

For the "if" part, suppose $\epsilon > 0$ is given, and P is a partition as in the statement of the theorem. Then Corollary 6.2 implies

$$0 \le U(f) - L(f) \le U(f; P) - L(f; P) < \epsilon.$$

Since ϵ is arbitrary, we conclude that $U(f) = L(f)$. □

The meaning of the previous theorem is that, a function is integrable exactly when the lower sum and the upper sum can be made arbitrarily close to each other by picking suitable partitions. This theorem is often easier to use than the original definition because it involves finding only *one* suitable partition. In the original definition of integrability, one has to consider *all* the partitions of the interval.

6.1.5   *Integrability of Continuous Functions*

We now illustrate the above ϵ-criterion by showing that continuous functions are integrable.

Theorem 6.3. *Suppose f is a continuous function on $[a,b]$. Then f is integrable on $[a,b]$.*

Ideas of Proof. We need to estimate

$$U(f;P) - L(f;P) = \sum_{i=1}^{n} (u_i(f) - l_i(f)) \Delta x_i.$$

The plan is to choose a partition P with sufficiently small norm such that each difference $(u_i(f) - l_i(f))$ is suitably small. The sum of the Δx_i is, in any case, $(b-a)$.

Proof. Suppose given $\epsilon > 0$. The function f is uniformly continuous on $[a,b]$ by Theorem 4.9. Thus, given $\epsilon_0 = \frac{\epsilon}{b-a}$, there exists $\delta > 0$ such that $|x - y| < \delta$ with $x, y \in [a,b]$ implies

$$|f(x) - f(y)| < \epsilon_0 = \frac{\epsilon}{b-a}.$$

Pick any partition P of $[a,b]$ with $\|P\| < \delta$; Exercise (3) below guarantees its existence. By the Extreme Value Theorem 4.6, both $u_i(f)$ and $l_i(f)$ are in the range of f on $[x_{i-1}, x_i]$, so

$$u_i(f) - l_i(f) < \epsilon_0$$

for each i. Thus, we have

$$U(f;P) - L(f;P) < \sum_{i=1}^{n} \epsilon_0 \Delta x_i = \epsilon_0(b-a) = \epsilon.$$

Theorem 6.2 now says that f is integrable on $[a,b]$. $\square$

In the exercises and later sections, we will see that a general integrable function may have many points of discontinuity.

6.1.6   *Exercises*

(1) If P is a partition of $[a,b]$, prove that $\Delta x_i \le \|P\|$ for each i and that $\sum_{i=1}^{n} \Delta x_i = b-a$.
(2) Suppose $P \subseteq Q$ are partitions of $[a,b]$. Prove that $\|Q\| \le \|P\| \le b-a$.
(3) Suppose $[a,b]$ is a closed interval, and $\delta > 0$. Prove that there exists a partition P of $[a,b]$ with $\|P\| < \delta$.
(4) Using Theorem 6.2, write down what it means for a bounded function f on $[a,b]$ to not be integrable on $[a,b]$.

(5) Suppose f is integrable on $[a, b]$, and there exists a real number r such that

$$L(f; P) \leq r \leq U(f; P)$$

for every partition P of $[a, b]$. Prove that $r = \int_a^b f$.

(6) Suppose f and g are bounded functions on $[a, b]$, and P is a partition on $[a, b]$. Prove that

$$u_i(f + g) \leq u_i(f) + u_i(g) \quad \text{and} \quad l_i(f + g) \geq l_i(f) + l_i(g)$$

for each i.

(7) In the proof of the second inequality in Theorem 6.1, an induction argument was used. Write down the details of this induction.

(8) Prove the last two inequalities in Theorem 6.1.

(9) Prove that every constant function is integrable on every closed bounded interval. Write down an expression for the integral.

(10) Prove that the step function J in Exercise (10) on page 104 is integrable on every closed bounded interval.

(11) Suppose f is an increasing function on $[a, b]$, and P is a partition of $[a, b]$. Prove that

$$U(f; P) - L(f; P) \leq \|P\|(f(b) - f(a)).$$

(12) Using the previous exercise or otherwise, prove that an increasing function on $[a, b]$ is integrable.

(13) Prove that every monotone function is integrable on $[a, b]$.

(14) Suppose n is a positive integer, and $[a_i, b_i] \subseteq [a, b]$ for $i = 1, \ldots, n$ are n sub-intervals such that $b_i \leq a_{i+1}$ for $1 \leq i \leq n - 1$. Prove that

$$\sum_{i=1}^{n} [U(f; [a_i, b_i]) - L(f; [a_i, b_i])] \leq (M - m)(b - a),$$

where $m = \inf\{f(x) : x \in [a, b]\}$ and $M = \sup\{f(x) : x \in [a, b]\}$.

(15) Using the previous exercise or otherwise, prove the following **Cauchy criterion for integrability**. A bounded function f on $[a, b]$ is integrable on $[a, b]$ if and only if, for each $\epsilon > 0$, there exists $\delta > 0$ such that

$$U(f; P) - L(f; P) < \epsilon$$

for *every* partition P of $[a, b]$ with $\|P\| < \delta$.

6.2 Integration via Tagged Partitions

There is another rigorous approach to integrals based on Riemann sums that is much closer to the usual presentation in calculus. In this section we discuss integrability based on Riemann sums. The main idea of Riemann sums is that, instead of taking the infimum or the supremum of f on each sub-interval, one takes the value of f at some point. Intuitively, this process forms a rectangle whose height lies between $l_i(f)$ and $u_i(f)$.

6.2.1   Riemann Sums

To define Riemann sums, first we need to define partitions with chosen points in each sub-interval.

Definition 6.4. A tagged partition

$$P = \{\{x_j\}_{j=0}^n, \{w_i\}_{i=1}^n\}$$

of an interval $[a, b]$ consists of:

- a partition $\{x_0 < \cdots < x_n\}$ of $[a, b]$ and
- a point w_i in $[x_{i-1}, x_i]$ for each $i = 1, \ldots, n$.

The **norm** of such a tagged partition is defined as

$$\|P\| = \max\{x_i - x_{i-1} : i = 1, \ldots, n\}.$$

Given a tagged partition P, one can forget about the points w_i and obtain a partition, which we also denote by P. Conversely, given a partition P of $[a, b]$, there are many different tagged partitions with the same partition points x_j that can be associated to it, which we denote by P', P'', etc.

Definition 6.5. Given a bounded function f on $[a, b]$ and a tagged partition $P = \{\{x_j\}_{j=0}^n, \{w_i\}_{i=1}^n\}$ of $[a, b]$, the **Riemann sum of f with respect to P** is the sum

$$R(f; P) = \sum_{i=1}^n f(w_i)\Delta x_i,$$

where $\Delta x_i = x_i - x_{i-1}$ as in Definition 6.1

Whenever we use the symbol $R(f; P)$, it is automatically assumed that P is a tagged partition. If the lower sum $L(f; P)$ or the upper sum $U(f; P)$ are also considered, then we are using the partition associated to the tagged partition P.

The following basic observation tells us how the lower sum, the upper sum, and the Riemann sum are related. The proof of this observation is left as an exercise.

Lemma 6.1. *Suppose f is a bounded function on $[a, b]$, and P is a tagged partition of $[a, b]$. Then*

$$L(f; P) \le R(f; P) \le U(f; P).$$

Therefore, for a given tagged partition, the Riemann sum always lies between the lower sum and the upper sum. Intuitively, this is true because in each sub-interval, a typical rectangle whose height is a value of f must lie between the inscribed and the circumscribed rectangles.

The next observation says that the lower sum and the upper sum can be closely approximated by Riemann sums.

Theorem 6.4. *Suppose f is a bounded function on $[a,b]$, P is a partition of $[a,b]$, and $\epsilon > 0$. Then there exist tagged partitions P' and P'' with the same partition points as P such that*

$$R(f;P') - L(f;P) < \epsilon \quad and \quad U(f;P) - R(f;P'') < \epsilon.$$

Ideas of Proof. The lower sum is formed using the infimum of f on each sub-interval. To construct the tagged partition P', in each sub-interval we choose a point whose value under f is very close to the infimum. The resulting Riemann sum is then close to the lower sum.

Proof. Set $\epsilon_0 = \frac{\epsilon}{b-a}$. If $P = \{x_0 < \cdots < x_n\}$, then since

$$l_i(f) + \epsilon_0 > l_i(f)$$

for each i, there exists a point w_i in $[x_{i-1}, x_i]$ such that

$$l_i(f) \le f(w_i) < l_i(f) + \epsilon_0.$$

So we have

$$0 \le f(w_i) - l_i(f) < \epsilon_0.$$

Let P' be the tagged partition with chosen points w_i and the same partition points as P. Multiplying by Δx_i and summing over the sub-intervals, we obtain the inequalities

$$R(f;P') - L(f;P) = \sum_{i=1}^{n}[f(w_i) - l_i(f)]\Delta x_i < \epsilon_0 \sum_{i=1}^{n} \Delta x_i = \epsilon_0(b-a) = \epsilon.$$

This proves the first inequality. The proof of the other inequality is left as an exercise. $\qquad\square$

6.2.2 Riemann Integrability

We now define a version of integrability based on Riemann sums.

Definition 6.6. Suppose f is a bounded function on $[a,b]$. Then f is said to be **Riemann integrable** on $[a,b]$ if there exists a real number $\mathcal{R}\int_a^b f$ such that, for each $\epsilon > 0$, there exists $\delta > 0$ such that

$$\left| R(f;P) - \mathcal{R}\int_a^b f \right| < \epsilon$$

for every tagged partition P with $\|P\| < \delta$. In this case, the real number $\mathcal{R}\int_a^b f$ is called the **Riemann integral** of f on $[a,b]$.

Note that we are using the notation $\mathcal{R}\int_a^b f$ to distinguish it from the integral $\int_a^b f$ in Definition 6.3 because at this moment we have not shown that they are equal. We also have not shown that integrability is equivalent to Riemann integrability. We now show that the two approaches to integration are equivalent.

Theorem 6.5. *A bounded function f on $[a,b]$ is integrable if and only if it is Riemann integrable. In this case, the integral $\int_a^b f$ and the Riemann integral $\mathcal{R}\int_a^b f$ are equal.*

Ideas of Proof. Under the assumption of integrability, the integral $\int_a^b f$ can be closely approximated by lower and upper sums, which in turn can be closely approximated by Riemann sums. Likewise, with Riemann integrability, the Riemann integral $\mathcal{R}\int_a^b f$ can also be closely approximated by Riemann sums. Therefore, it should be intuitively clear that the two approaches are equivalent. Also, it should not be surprising that some sort of $\frac{\epsilon}{2}$-argument is involved as we link the integral to the Riemann integral via lower, upper, and Riemann sums.

Proof. For the "only if" part, we assume f is integrable on $[a,b]$, and suppose $\epsilon > 0$. By Exercise (15) on page 141, there exists $\delta > 0$ such that

$$U(f;P) - L(f;P) < \epsilon \tag{6.3}$$

for every partition P with $\|P\| < \delta$. Now if P is any tagged partition, then

$$L(f;P) \le R(f;P) \le U(f;P) \quad \text{and} \quad L(f;P) \le \int_a^b f \le U(f;P)$$

by Lemma 6.1 and integrability. Therefore, if P has norm $< \delta$, then (6.3) implies

$$\left| R(f;P) - \int_a^b f \right| < \epsilon.$$

This shows that f is Riemann integrable and $\int_a^b f = \mathcal{R}\int_a^b f$.

For the "if" part, we assume f is Riemann integrable. We will use the ϵ-criterion in Theorem 6.2 and an $\frac{\epsilon}{4}$-argument to show that f is integrable. Given $\epsilon > 0$, Riemann integrability implies that there exists $\delta > 0$ such that

$$\left| R(f;P) - \mathcal{R}\int_a^b f \right| < \frac{\epsilon}{4}$$

for every tagged partition P with norm $< \delta$. Pick any partition P with norm $< \delta$. By Theorem 6.4, there exist tagged partitions P' and P'' with the same partition points as P, and hence with norms $< \delta$, such that

$$R(f;P') - L(f;P) < \frac{\epsilon}{4} \quad \text{and} \quad U(f;P) - R(f;P'') < \frac{\epsilon}{4}.$$

Since we can write

$$U(f;P) - L(f;P) = [U(f;P) - R(f;P'')] + \left[R(f;P'') - \mathcal{R}\int_a^b f\right]$$

$$+ \left[\mathcal{R}\int_a^b f - R(f;P')\right] + [R(f;P') - L(f;P)],$$

the previous three inequalities and the Triangle Inequality imply

$$|U(f;P) - L(f;P)| < 4 \cdot \frac{\epsilon}{4} = \epsilon. \tag{6.4}$$

Theorem 6.2 now says that f is integrable.

To see that the integral is equal to the Riemann integral in this case, we use the same inequalities in the previous paragraph as follows.

$$U(f;P) < R(f;P'') + \frac{\epsilon}{4} < \mathcal{R}\int_a^b f + \frac{\epsilon}{2}$$

$$< R(f;P') + \frac{3\epsilon}{4} < L(f;P) + \epsilon \leq U(f;P) + \epsilon.$$

Therefore, we have

$$\left| \mathcal{R} \int_a^b f - U(f;P) \right| < \frac{\epsilon}{2}.$$

But we also have

$$\left| U(f;P) - \int_a^b f \right| < \epsilon$$

by (6.4). Combining the previous two inequalities, we obtain

$$\left| \mathcal{R}\int_a^b f - \int_a^b f \right| \leq \left| \mathcal{R}\int_a^b f - U(f;P) \right| + \left| U(f;P) - \int_a^b f \right| < \frac{3\epsilon}{2}$$

Since $\epsilon > 0$ is arbitrary, we conclude that $\mathcal{R}\int_a^b f = \int_a^b f$. $\qquad\square$

In view of Theorem 6.5, the words *integrable* and *Riemann integrable* are interchangeable.

6.2.3   *Exercise*

(1) Using Definition 6.6, write down what it means for a bounded function f on $[a,b]$ to not be Riemann integrable on $[a,b]$.
(2) Prove Lemma 6.1.
(3) Prove the second inequality in Theorem 6.4.
(4) Without using Theorem 6.5, prove that when f is Riemann integrable on $[a,b]$, the value of the Riemann integral $\mathcal{R}\int_a^b f$ is unique.
(5) Prove the following **Cauchy criterion for Riemann integrability**. A bounded function f on $[a,b]$ is Riemann integrable if and only if, for every $\epsilon > 0$, there exists $\delta > 0$ such that

$$|R(f;P) - R(f;Q)| < \epsilon$$

for any two tagged partitions P and Q with norms $< \delta$.
(6) Using the previous exercise, prove that the characteristic function $\chi_{\mathbb{Q}}$ of $\mathbb{Q}$ is not Riemann integrable on $[0,1]$.

(7) Suppose f is Riemann integrable on $[a,b]$ and that each P_n is a tagged partition with $\lim \|P_n\| = 0$. Prove that

$$\mathcal{R} \int_a^b f = \lim_{n \to \infty} R(f; P_n).$$

(8) For an integrable function f on $[0,1]$, prove that

$$\int_0^1 f = \lim_{n \to \infty} \sum_{i=1}^n f\left(\frac{i}{n}\right) \frac{1}{n} = \lim_{n \to \infty} \sum_{i=1}^n f\left(\frac{i-1}{n}\right) \frac{1}{n}.$$

(9) Give an example of a bounded function f for which the limits in the previous exercise both exist, but f is not integrable on $[0,1]$.

(10) Give an example of an integrable function f on $[0,1]$ such that $\int_0^1 f = 0$ and that $f(x) \neq 0$ for all x in $[0,1]$.

(11) Prove that the function $f:[0,1] \to \mathbb{R}$ given by $f(x) = \sin\left(\frac{1}{x}\right)$ if $x \neq 0$ and $f(0) = 0$ is integrable.

(12) Consider the function $f:\mathbb{R} \to \mathbb{R}$ given by $f(x) = x$ if x is rational and $f(x) = -x$ if x is irrational. Prove that f is not integrable on any closed bounded interval.

6.3 Basic Properties of Integrals

In this section we establish some basic properties of the integral, most of which should already be familiar to the reader from calculus.

6.3.1 Linearity

The next result says that integrals are linear and respect inequalities.

Theorem 6.6. *Suppose f and g are integrable on $[a,b]$, and r is a real number.*

(1) The sum $f + g$ is integrable on $[a,b]$, and

$$\int_a^b (f+g) = \int_a^b f + \int_a^b g.$$

(2) The scalar multiple rf is integrable on $[a,b]$, and

$$\int_a^b rf = r \int_a^b f.$$

(3) If $f(x) \geq 0$ for all x in $[a,b]$, then

$$\int_a^b f \geq 0.$$

Proof.   We will prove the first part. The other two parts are left as exercises. Suppose P_1 and P_2 are partitions of $[a,b]$ and $P = P_1 \cup P_2$. Then we have inequalities

$$L(f; P_1) + L(g; P_2) \leq L(f; P) + L(g; P) \leq L(f+g; P) \leq U(f+g; P)$$
$$\leq U(f; P) + U(g; P) \leq U(f; P_1) + U(g; P_2).$$

The first, the third, and the fifth inequalities follow from Theorem 6.1, while the second and the fourth inequalities follow from Exercise (6) on page 141. Using the above inequalities and Corollary 6.2, we obtain the inequalities

$$L(f; P_1) + L(g; P_2) \le L(f + g) \le U(f + g) \le U(f; P_1) + U(g; P_2).$$

Since this holds for arbitrary partitions P_1 and P_2, it follows that

$$\int_a^b f + \int_a^b g \le L(f + g) \le U(f + g) \le \int_a^b f + \int_a^b g,$$

from which the first part follows.   □

Corollary 6.3. *Suppose f and g are integrable on* $[a, b]$ *such that*

$$f(x) \ge g(x)$$

for all x in $[a, b]$. *Then*

$$\int_a^b f \ge \int_a^b g.$$

Proof.   By Theorem 6.6, the difference $f - g$ is integrable, and

$$0 \le \int_a^b (f - g) = \int_a^b f - \int_a^b g,$$

from which the desired inequality follows.   □

6.3.2   *Consecutive Intervals*

The next result says that integrals can be added over two consecutive intervals.

Theorem 6.7. *Suppose f is integrable on* $[a, c]$ *and on* $[c, b]$. *Then f is integrable on* $[a, b]$ *and*

$$\int_a^b f = \int_a^c f + \int_c^b f.$$

Ideas of Proof. Partitions of $[a, c]$ and $[c, b]$ can be spliced together to form a partition of $[a, b]$. This implies that the lower and the upper sums also have the same property. Therefore, close estimates between the lower and the upper sums on $[a, c]$ and $[c, b]$ can be extended to all of $[a, b]$.

Proof.   This is an $\frac{\epsilon}{2}$-argument. Given $\epsilon > 0$, there exist partitions P_1 of $[a, c]$ and P_2 of $[c, b]$ such that

$$U(f; P_i) - L(f; P_i) < \frac{\epsilon}{2}$$

for each i. The union $P = P_1 \cup P_2$ is a partition of $[a, b]$, and

$$U(f; P) - L(f; P) = \sum_{i=1}^2 (U(f; P_i) - L(f; P_i)) < \frac{\epsilon}{2} + \frac{\epsilon}{2} = \epsilon.$$

Therefore, f is integrable on $[a, b]$. To prove the desired equality, consider the inequalities

$$\int_a^c f + \int_c^b f - \epsilon \le U(f; P_1) + U(f; P_2) - \epsilon = U(f; P) - \epsilon$$

$$< L(f; P) \le \int_a^b f \le U(f; P) < L(f; P) + \epsilon$$

$$= L(f; P_1) + L(f; P_2) + \epsilon \le \int_a^c f + \int_c^b f + \epsilon.$$

The desired equality $\int_a^b f = \int_a^c f + \int_c^b f$ now follows because ϵ is arbitrary.   □

Since we defined $\int_b^a f = -\int_a^b f$ if f is integrable on $[a, b]$, the previous theorem is actually valid for all a, b, and c, as long as the three integrals exist.

6.3.3   *Mean Value Theorem for Integrals*

Intuitively, this result says that the integral of a continuous function is actually the area of a rectangle, whose width is that of the interval and whose height is the value of the function at a certain point.

Theorem 6.8. *Suppose f is continuous on $[a, b]$. Then there exists a point c in $[a, b]$ such that*

$$\int_a^b f = f(c)(b - a).$$

Proof.   Theorem 6.3 says that f is integrable. The result is obvious if f is a constant function, so we assume that f is not constant. We are trying to show that the real number

$$r = \frac{1}{b - a} \int_a^b f$$

is in the range of f. By the Extreme Value Theorem 4.6, f achieves its maximum and minimum on $[a, b]$ at some points β and α. If r is either $f(\alpha)$ or $f(\beta)$, then we are done. Otherwise, by the Intermediate Value Theorem 4.7 applied to the interval with end points α and β, it is enough to observe that r lies between $f(\alpha)$ and $f(\beta)$. From Theorem 6.1 we know that

$$f(\alpha)(b - a) \le \int_a^b f \le f(\beta)(b - a),$$

from which we obtain the desired inequalities $f(\alpha) \le r \le f(\beta)$.   □

6.3.4   *Exercises*

(1) Finish the proof of Theorem 6.6 by proving the last two parts.
(2) Suppose f is integrable on $[a, b]$. Prove that $|f|$ is integrable on $[a, b]$ and

$$\left| \int_a^b f \right| \le \int_a^b |f|.$$

(3) Give an example where $|f|$ is integrable, but f is not.

(4) Suppose f is integrable on $[a, b]$ and $[c, d] \subseteq [a, b]$. Prove that f is integrable on $[c, d]$.

(5) Suppose f is integrable on $[a, b]$, g is a bounded function on $[a, b]$, and that $f(x) = g(x)$ on $[a, b]$ except for finitely many points. Prove that g is integrable on $[a, b]$ and $\int_a^b f = \int_a^b g$.

(6) Suppose f is continuous on $[a, b]$ such that $\int_a^b f = 0$. Prove that $f(c) = 0$ for some point c in $[a, b]$.

(7) Suppose f and g are continuous on $[a, b]$ such that $\int_a^b f = \int_a^b g$. Prove that $f(c) = g(c)$ for some point c in $[a, b]$.

(8) Suppose f is continuous on $[a, b]$ such that $f(x) \geq 0$ for all x and $\int_a^b f = 0$. Prove that $f(x) = 0$ for all x.

(9) Suppose f is integrable on $[a, b]$, g is continuous on $[c, d]$, and that $f([a, b]) \subseteq [c, d]$. Prove that the composition $g \circ f$ is integrable on $[a, b]$.

(10) Suppose f is integrable on $[a, b]$. Prove that the product f^2 is integrable on $[a, b]$.

(11) Suppose f and g are integrable on $[a, b]$. Prove that the product fg is integrable on $[a, b]$.

(12) Suppose f and g are continuous on $[a, b]$ and $g(x) > 0$ for all x. Prove that there exists a point c in $[a, b]$ such that

$$\int_a^b fg = f(c) \int_a^b g.$$

Use this result to give another proof of Theorem 6.8.

(13) Give an example to show that in Theorem 6.8, the continuity assumption cannot be weakened to integrability.

(14) Consider the function on $[0, 1]$ given by $f(x) = 0$ if x is irrational, $f(0) = 1$, and $f(\frac{m}{n}) = \frac{1}{n}$ if $\frac{m}{n}$ is a rational number in lowest terms. Prove the following statements.

    (a) f is continuous at each irrational number.
    (b) f is discontinuous at each rational number.
    (c) f is integrable.

This function is called the **Thomae function.**

6.4 Fundamental Theorem of Calculus

The main purpose of this section is to prove the theorem in the section title, which the reader should already be familiar with from calculus. Some important consequences are then discussed. As in the previous sections, our goal here is to prove these theorems rigorously, rather than doing examples about how to use them as in calculus.

6.4.1 *Motivation*

Just like chain rule is the most important derivative rule in differential calculus, the Fundamental Theorem of Calculus is the most important result in integral calculus. Chain rule is important because a lot of functions are compositions of simpler functions. Computing the actual value of an integral is, however, an entirely different matter. It involves first computing the lower and upper sums with respect to *every* partition of the interval, and then taking the supremum of the lower sums and the infimum of the upper sums. Moreover, the integrals of the constituent functions do not usually tell us what the integral is for the composition. For example, try to compute $\int_0^1 x^2$, and see if it helps you compute $\int_0^1 x^4$ and $\int_0^1 x^6$ from the definition.

It is therefore necessary to develop a more convenient method for computing the integral than using the definition. The Fundamental Theorem of Calculus serves exactly this purpose. It drastically simplifies the computation of the integral for a function that is actually the derivative of another function.

6.4.2 *Main Theorem*

Here is the Fundamental Theorem of Calculus.

Theorem 6.9. *Suppose f is continuous on $[a,b]$, differentiable on (a,b), and that f' is integrable on $[a,b]$. Then*

$$\int_a^b f' = f(b) - f(a).$$

Ideas of Proof. We need to estimate

$$\left| \int_a^b f' - (f(b) - f(a)) \right|.$$

Since $\int_a^b f'$ can be closely estimated by the lower and the upper sums, we will try to do the same to the difference $(f(b) - f(a))$. Remember the Mean Value Theorem 5.7 relates such a difference to the derivative, so it is no surprise that we will use this theorem.

Proof. Given $\epsilon > 0$, by Theorem 6.2, there exists a partition $P = \{x_0 < \cdots < x_n\}$ of $[a,b]$ such that

$$U(f'; P) - L(f'; P) < \epsilon.$$

Since ϵ is arbitrary and

$$L(f'; P) \le \int_a^b f' \le U(f'; P),$$

it is enough to show that $(f(b) - f(a))$ satisfies the same inequalities, that is,

$$L(f'; P) \le f(b) - f(a) \le U(f'; P). \tag{6.5}$$

Applying the Mean Value Theorem to each sub-interval $[x_{i-1}, x_i]$, we obtain a point w_i in (x_{i-1}, x_i) such that

$$f'(w_i)\Delta x_i = f(x_i) - f(x_{i-1}).$$

Since

$$l_i(f'; P) \le f'(w_i) \le u_i(f'; P),$$

we conclude that

$$L(f'; P) = \sum_{i=1}^{n} l_i(f'; P)\Delta x_i \le \sum_{i=1}^{n} (f(x_i) - f(x_{i-1}))$$
$$\le \sum_{i=1}^{n} u_i(f'; P)\Delta x_i = U(f'; P).$$

The inequalities (6.5) now follow because

$$\sum_{i=1}^{n} (f(x_i) - f(x_{i-1})) = f(b) - f(a).$$

□

Using the Fundamental Theorem of Calculus to compute integrals is something the reader should already know from calculus. We adopt the notation

$$f(b) - f(a) = f \Big|_a^b$$

from calculus. We now consider other important integration theorems.

6.4.3   *Integration by Parts*

This theorem is the integral form of the product rule (Theorem 5.2), which says

$$(fg)' = f'g + fg'.$$

Roughly speaking, integrating both sides from here yields Integration by Parts. The formal proof uses the Fundamental Theorem of Calculus.

Corollary 6.4. *Suppose f and g are continuous on $[a, b]$, differentiable on (a, b), and that f' and g' are integrable on $[a, b]$. Then*

$$\int_a^b fg' = fg \Big|_a^b - \int_a^b f'g.$$

Proof.   The desired equality is exactly that in the Fundamental Theorem of Calculus 6.9 for the product $h = fg$, which satisfies the hypotheses by Exercise (3) on page 93, Theorem 5.2, Theorem 6.3, and Exercise (11) on page 149.   □

6.4.4   Integral Form of Taylor's Theorem

In Taylor's Theorem 5.8, it was shown that for $n \geq 1$ the error term

$$R_n(x; a) = f(x) - T_n(x; a)$$

in the degree n Taylor polynomial approximation of f at a takes the form

$$\frac{f^{(n+1)}(c)}{(n+1)!}(x-a)^{n+1}$$

for some point c between a and x. We now present another form of this error term that does not give preference to any particular point in the interval determined by a and x. To do that, we will have to use the integral. This result is also a good demonstration of the Fundamental Theorem of Calculus and Integration by Parts. We want to keep using x as the independent variable in the integral, so the point x in the error term will be called b below.

Theorem 6.10. *Suppose $f : I = [\alpha, \beta] \to \mathbb{R}$ is a function such that $f^{(k)}$ exists on (α, β) and is continuous on I for $0 \leq k \leq n+1$. Then given two distinct points a and b in (α, β), we have*

$$R_n(b; a) = \frac{1}{n!} \int_a^b (b-x)^n f^{(n+1)}(x)\,dx.$$

Proof.   This is an induction proof for $k = 1, 2, \ldots, n$. The kth case is

$$R_k(b; a) = \frac{1}{k!} \int_a^b (b-x)^k f^{(k+1)}, \tag{6.6}$$

so the first case is the assertion

$$R_1(b; a) = f(b) - [f(a) + f'(a)(b-a)] = \int_a^b (b-x)f''. \tag{6.7}$$

To prove the first case, we use the Fundamental Theorem of Calculus 6.9 and Integration by Parts with $g = x - b$. Since $g' = 1$, the middle expression in (6.7) can be rewritten as

$$\int_a^b (f' - f'(a)) = (f' - f'(a))g \Big|_a^b - \int_a^b (x-b)f'' = \int_a^b (b-x)f''$$

because

$$(f' - f'(a))(a) = 0 = g(b).$$

This proves the first case of (6.6). The induction step is a similar argument using Integration by Parts, which is left as an exercise.   □

6.4.5 *Derivative of the Integral*

In many calculus textbooks, the next two results are often collectively called the first Fundamental Theorem of Calculus. The theme here is that a function defined by an integral is a little bit nicer than the function being integrated.

Theorem 6.11. *Suppose f is integrable on $[a,b]$. Define a function on $[a,b]$ by*

$$F(x) = \int_a^x f.$$

Then F is uniformly continuous on $[a,b]$.

Ideas of Proof. The function F can be intuitively interpreted as the area under f over $[a,x]$. As x moves within the interval $[a,b]$, the values of F vary nicely because f is integrable. For the next two proofs, we need to estimate

$$F(x) - F(y) = \int_a^x f - \int_a^y f = \int_y^x f.$$

We can do this by estimating f and $|x - y|$.

Proof. Since f is bounded, there exists $M > 0$ such that

$$|f(x)| \le M \quad \text{for all} \quad x \in [a,b].$$

Given $\epsilon > 0$, let $\delta = \frac{\epsilon}{M}$. Then for $x > y$ in $[a,b]$ with $x - y < \delta$, we have

$$|F(x) - F(y)| \le \int_y^x |f| \le M(x - y) < \epsilon.$$

We have a similar inequality when $x < y$. This shows that F is uniformly continuous on $[a,b]$. □

Theorem 6.12. *Suppose f and F are as in Theorem 6.11 and that f is continuous at a point c in (a,b). Then F is differentiable at c and*

$$F'(c) = f(c).$$

Proof. First observe that

$$F(x) - F(c) = \int_c^x f \quad \text{and} \quad f(c) = \frac{1}{x - c} \int_c^x f(c)$$

for $x \ne c$. Given $\epsilon > 0$, the continuity of f at c implies that there exists $\delta > 0$ such that

$$|x - c| < \delta \text{ with } x \in [a,b] \quad \text{implies} \quad |f(x) - f(c)| < \epsilon.$$

Therefore, for such x we have

$$\left| \frac{F(x) - F(c)}{x - c} - f(c) \right| = \left| \frac{1}{x - c} \int_c^x (f - f(c)) \right| < \epsilon,$$

which proves $F'(c) = f(c)$ by Theorem 4.3. □

6.4.6   *Substitution*

The next result is often called u-substitution or change of variables in calculus textbooks.

Theorem 6.13. *Suppose g is a continuously differentiable function on an open interval I, J is an open interval containing the image of g, and f is a continuous function on J. Then for any two points $a < b$ in I, we have*

$$\int_a^b f(g(x))g'(x)dx = \int_{g(a)}^{g(b)} f(u)du$$

Proof.   First note that the composition $f(g(x))$ is continuous on I, and therefore so is the product $f(g(x))g'(x)$. It is integrable on $[a,b]$ by Theorem 6.3. Define

$$F(x) = \int_{g(a)}^x f$$

for x in $g([a,b])$. It is differentiable with $F'(c) = f(c)$ by Theorem 6.12. Chain rule yields

$$[F(g(x))]' = F'(g(x))g'(x) = f(g(x))g'(x).$$

Therefore, it follows from Theorem 6.9 that

$$\int_a^b f(g(x))g'(x)dx = \int_a^b (F \circ g)' = F \circ g \Big|_a^b = \int_{g(a)}^{g(b)} f,$$

as desired.   □

6.4.7   *Exercises*

(1) Suppose f is continuous on $[a,b]$ such that $\int_a^x f = \int_x^b f$ for all x in $[a,b]$. Prove that f is the constant function 0.

(2) For a continuous function f on $\mathbb{R}$, define $g(x) = \int_{x-1}^{x+1} f$. Prove that g is differentiable on $\mathbb{R}$. Then compute g'.

(3) Write down the details of the proof of Corollary 6.4.

(4) Finish the induction step in the proof of Theorem 6.10 as follows.

(a) First prove that

$$R_{k+1}(b;a) = R_k(b;a) - \frac{f^{(k+1)}(a)}{(k+1)!}(b-a)^{k+1}.$$

(b) Suppose the kth case (6.6) is true. Prove that

$$R_{k+1}(b;a) = \frac{1}{k!}\int_a^b [f^{(k+1)} - f^{(k+1)}(a)](b-x)^k.$$

(c) Apply Integration by Parts to the previous part to prove the $(k+1)$st case of (6.6).

(5) Under the hypotheses of Theorem 6.10, suppose further that $a < b$ and that $m \leq |f^{(n+1)}| \leq M$ on (α, β). Prove that
$$\frac{m(b-a)^{n+1}}{(n+1)!} \leq R_n(b;a) \leq \frac{M(b-a)^{n+1}}{(n+1)!}.$$

(6) Using Theorem 6.10 with $a < b$, give another proof of the original Taylor's Theorem 5.8, that is,
$$R_n(b;a) = \frac{f^{(n+1)}(c)}{(n+1)!}(b-a)^{n+1}$$
for some point c with $a < c < b$.

(7) Write down the details of the proof of Theorem 6.11 in the case $x < y$.

(8) In the proof of Theorem 6.12, we used the ϵ-δ characterization of the limit $\lim_{x \to c} \frac{F(x)-F(c)}{x-c}$. Write down a version of that proof that uses the sequential definition of limits (Definition 4.2).

6.5 Additional Exercises

(1) Suppose f is continuous on $[0, \infty)$, $f(x) > 0$ for all $x > 0$, and that $[f(x)]^n = n \int_0^x f^{n-1}$ for some integer $n \geq 2$. Prove that $f(x) = x$ for all x. Here f^{n-1} is the product of $n-1$ copies of f.

(2) Suppose f and g are integrable on $[a,b]$. Prove that $\min\{f,g\}$ and $\max\{f,g\}$ are both integrable on $[a,b]$.

(3) Prove that every function of bounded variation is integrable on a closed bounded interval.

(4) Suppose f and g are integrable on some intervals, that they are both monotone, and that the composition $f \circ g$ is defined. Prove that $f \circ g$ is integrable on the domain of g.

(5) Prove that the function $f(x) = 1$ if $x = \frac{1}{n}$ for $n \in \mathbb{Z}_+$ and $f(x) = 0$ otherwise is integrable on $[0,1]$.

(6) A function f on $[a,b]$ is called **piecewise continuous** if there exists a partition $P = \{x_0 < \cdots < x_n\}$ of $[a,b]$ such that f is uniformly continuous on each open sub-interval (x_{i-1}, x_i). Prove the following statements.

   (a) A piecewise continuous function on $[a,b]$ is bounded.
   (b) A piecewise continuous function on $[a,b]$ is integrable on $[a,b]$.

(7) A function f on $[a,b]$ is called **piecewise monotone** if there exists a partition $P = \{x_0 < \cdots < x_n\}$ of $[a,b]$ such that f is monotone on each open sub-interval (x_{i-1}, x_i). Prove that a bounded piecewise monotone function on $[a,b]$ is integrable on $[a,b]$.

(8) Suppose f is bounded on $[a,b]$. Prove that f is integrable on $[a,b]$ if and only if it is integrable on each closed sub-interval of (a,b). In this case, find an expression of $\int_a^b f$ in terms of the integrals of f over closed sub-intervals of (a,b).

(9) Suppose f is a bounded function on $[a,b]$ such that $\{R(f;P_n)\}$ is a convergent sequence whenever $\{P_n\}$ is a sequence of tagged partitions with $\lim\|P_n\| = 0$. Prove that f is integrable on $[a,b]$.

(10) If n is a positive integer, give an example of an integrable function on some closed bounded interval that has exactly n points of discontinuity.

(11) Give an example of an increasing integrable function on some closed bounded interval that has countably many points of discontinuity.

(12) Give an example of an integrable function f on $[a,b]$ and a integrable function g on $[c,d]$ such that $f([a,b]) \subseteq [c,d]$ and that the composition $g \circ f$ is not integrable on $[a,b]$.

(13) Suppose f is defined on $[a,\infty)$ and is integrable on $[a,b]$ for all $b > a$. Define the **improper integral** $\int_a^\infty f = \lim_{b\to\infty} \int_a^b f$, if the limit exists. In this case, we say the improper integral converges. Prove the **Integral Test** as follows. Suppose $\sum a_n$ is a series with $\{a_n\}$ decreasing and non-negative. Suppose f is a non-negative decreasing function on $[1,\infty)$ such that $f(n) = a_n$ for each n.

(a) Prove that

$$\sum_{i=2}^n a_i = L(f;P) \le \int_1^n f \le U(f;P) = \sum_{i=1}^{n-1} a_i$$

for the partition $P = \{1 < 2 < \cdots < n\}$ of $[1,n]$.

(b) Prove that the series $\sum a_n$ converges if and only if the improper integral $\int_1^\infty f$ converges.

Chapter 7

Sequences and Series of Functions

In this chapter, we discuss sequences and series of functions and their convergence behavior. One main reason for studying sequences of functions is that it is a very convenient way to construct examples of functions with desired properties. Two concepts of convergence, pointwise and uniform, of sequences of functions are introduced in section 7.1. In section 7.2 we discuss properties that are sometimes preserved by the limits of a sequence of functions. Series of functions, which are a particular kind of sequences of functions, are discussed in section 7.3. Power series, which are a particular kind of series of functions with very nice properties, are discussed in section 7.4.

7.1   Pointwise and Uniform Convergence

In this section, we first introduce pointwise convergence of a sequence of functions. It is the most straight forward way to define convergence in this setting. Several examples are then given to illustrate the inadequacy of this form of convergence. Then we introduce a stronger form of convergence called uniform convergence. In the next section, we will show that uniform convergence allows the passage of properties to the limits.

7.1.1   *Pointwise Convergence*

Throughout this section, unless otherwise specified, $\{f_n\}$ denotes a sequence of functions $f_1, f_2, \ldots$, all having the same domain $S \subseteq \mathbb{R}$. Unlike a sequence of real numbers as in Definition 2.1, there are now *three* ways to consider limits of $\{f_n(x)\}$. First, we can consider limits with respect to the sequential index n for a fixed point x in S. Second, we can consider limits of $f_n(x)$ with respect to the points $x \in S$ for a fixed index n. Finally, we can try to vary both n and x in $f_n(x)$. We begin with the first version of convergence.

Definition 7.1. Let $\{f_n\}$ be a sequence of functions and f be a function defined on $S \subseteq \mathbb{R}$. Then we say the sequence **converges pointwise to f on** S if, for each x in

S, the sequence $\{f_n(x)\}$ converges to $f(x)$. In this case, we say f is the **pointwise limit** of the sequence on S and write

$$f_n \to f \text{ pointwise } \quad \text{or} \quad f = \lim_{n \to \infty} f_n.$$

So $f_n \to f$ pointwise on S means, for each x in S and each $\epsilon > 0$, there exists N such that

$$n \geq N \quad \text{implies} \quad |f_n(x) - f(x)| < \epsilon.$$

The key point here is that this N depends on *both* x and ϵ. In other words, even for a fixed $\epsilon > 0$, two different points x and y would normally require two different integers, say $N(x)$ and $N(y)$. As x varies through the domain S, the set of $N(x)$ may well be unbounded. Thus, there may not be a single integer N that can work for all the points x in S.

To see whether pointwise convergence is a good enough concept of convergence of sequences of functions, let us consider some examples. We will examine what properties, if any, are automatically transferred from the functions f_n to the pointwise limit. The reader should try to fill in the details of the following examples.

7.1.2   *Examples of Pointwise Convergence*

The first example demonstrates that the property of having a certain limit at a point is not usually transferred to the pointwise limit.

Example 7.1. Consider $f_n = x^n$ on $[0,1)$. Then $f_n \to f = 0$ pointwise on $[0,1)$. The point 1 is a limit point of $[0,1)$. But for each n we have $\lim_{x \to 1} f_n = 1$, while $\lim_{x \to 1} f = 0$. Therefore, we have

$$\lim_{x \to 1} \lim_{n \to \infty} f_n = 0 \neq 1 = \lim_{n \to \infty} \lim_{x \to 1} f_n. \qquad (7.1)$$

So the two limit operations, $\lim_{x \to 1}$ and $\lim_{n \to \infty}$, do not commute. This is another way of saying that the property of having limit 1 as $x \to 1$ is not transferred to the pointwise limit. To say it yet another way, preservation of the limit operation $\lim_{x \to c}$ is really about the commutation of $\lim_{x \to c}$ and $\lim_{n \to \infty}$.

The next example shows that continuity is not usually transferred to the pointwise limit.

Example 7.2. Modify the previous example, and consider $f_n = x^n$ on $[0,1]$. Define f on $[0,1]$ by $f(x) = 0$ if $0 \leq x < 1$ and $f(1) = 1$. Then $f_n \to f$ pointwise on $[0,1]$. Each f_n is continuous at 1, but f is not because

$$\lim_{x \to 1} \lim_{n \to \infty} f_n = \lim_{x \to 1} f(x) = 0 \neq 1 = f(1) = \lim_{n \to \infty} \lim_{x \to 1} f_n.$$

Therefore, we again have the non-commutation of limit operations, this time expressing the non-preservation of continuity in the pointwise limit.

The next example shows that integrals do not behave nicely with respect to pointwise convergence.

Example 7.3. For $n \geq 2$ consider the functions f_n on $[0,1]$ defined as

$$f_n(x) = \begin{cases} 0 & \text{if } 0 \leq x \leq 1 - \frac{2}{n}, \\ n^2(x-1) + 2n & \text{if } 1 - \frac{2}{n} \leq x \leq 1 - \frac{1}{n}, \\ -n^2(x-1) & \text{if } 1 - \frac{1}{n} \leq x \leq 1. \end{cases}$$

The easiest way to understand f_n is to look at its graph, which has a triangular spike over $[1 - \frac{2}{n}, 1]$ with height n. Each f_n is continuous, and hence integrable, on $[0,1]$ with $\int_0^1 f_n = 1$. On the other hand, we have $f_n \to f = 0$ pointwise on $[0,1]$ and $\int_0^1 f = 0$. Therefore, we have

$$\int_0^1 \lim_{n\to\infty} f_n = 0 \neq 1 = \lim_{n\to\infty} \int_0^1 f_n.$$

This non-commutation of the limit operation $\lim_{n\to\infty}$ and the integral $\int_0^1$ says that the integral of the pointwise limit does not need to be the limit of the sequence of integrals.

The following example illustrates that derivatives do not behave nicely with respect to pointwise convergence.

Example 7.4. Consider the functions $f_n = \frac{1}{n}\sin(n^2 x)$ on $[0,1]$. Then $f_n'(x) = n\cos(n^2 x)$, so $\lim_{n\to\infty} f_n'(x)$ does not exist for any x. On the other hand, we have $f_n \to f = 0$ pointwise on $[0,1]$, so $f' = 0$. Therefore, we have

$$\left[\lim_{n\to\infty} f_n(x)\right]' = 0 \neq \lim_{n\to\infty} f_n'(x).$$

This non-commutation of the limit operation $\lim_{n\to\infty}$ and the derivative says that the derivative of the pointwise limit does not need to be the limit of the sequence of derivatives.

As the examples above illustrate, pointwise convergence is not strong enough for the pointwise limit to inherit good properties from the functions f_n. We now introduce a stronger form of convergence, which will be studied further in the next section, that ensures that certain properties are preserved in the limit.

7.1.3 Uniform Convergence

Definition 7.2. Let $\{f_n\}$ be a sequence of functions and f be a function defined on $S \subseteq \mathbb{R}$. Then we say the sequence **converges uniformly to f on S** if, for each $\epsilon > 0$, there exists an integer N such that

$$n \geq N \quad \text{implies} \quad |f_n(x) - f(x)| < \epsilon \quad \text{for all } x \text{ in } S.$$

In this case, we say f is the **uniform limit** of the sequence on S and write $f_n \to f$ uniformly.

It is an easy exercise to see that uniform convergence implies pointwise convergence. Therefore, when $f_n \to f$ uniformly on S, we can also write $f = \lim_{n\to\infty} f_n$. The key point of uniform convergence is that, given $\epsilon > 0$, there is one integer N that works for all x in S. Intuitively, this says that the graphs of f_n for all n sufficiently large are within an ϵ-tube of the graph of f. We will show in the next section that uniform convergence forces the passage of several good properties to the uniform limit.

7.1.4 *Examples of Uniform Convergence*

Example 7.5. In Example 7.1, the sequence $f_n = x^n$ does not convergence uniformly to $f = 0$ on $[0,1)$. Indeed, set $\epsilon = \frac{1}{2}$. Then for any integer $N > 0$, any point x satisfying $2^{-\frac{1}{N}} \le x < 1$ yields

$$|x^N - 0| \ge \frac{1}{2} = \epsilon.$$

This shows that $f_n \to f$ pointwise but not uniformly on $[0,1)$.

Example 7.6. In Example 7.4, the sequence $f_n = \frac{1}{n}\sin(n^2 x)$ converges uniformly to $f = 0$ on $[0,1]$. Indeed, given $\epsilon > 0$, simply choose N such that $N\epsilon > 1$. Then for $n \ge N$ and all x in $[0,1]$, we have

$$\left|\frac{1}{n}\sin(n^2 x) - 0\right| \le \frac{1}{n} \le \frac{1}{N} < \epsilon.$$

Therefore, we have $f_n \to f$ uniformly on $[0,1]$. Example 7.4 and this example together imply that uniform convergence is still not strong enough to ensure the preservation of derivative in the uniform limit.

Example 7.7. On $[0,1]$ consider the function $f_n = \frac{x}{n}$. Then we have $f_n \to f = 0$ uniformly on $[0,1]$. Moreover, since $f_n' = \frac{1}{n}$, we have

$$\left[\lim_{n\to\infty} f_n(x)\right]' = 0 = \lim_{n\to\infty} f_n'(x),$$

unlike Example 7.4. Therefore, in this case the limit operation $\lim_{n\to\infty}$ and the derivative do commute.

7.1.5 *Uniformly Cauchy*

Definition 7.2 can be hard to use in practice because one has to know the uniform limit f in advance to show that f_n converges to it uniformly. Analogous to Theorem 2.10, we would like to describe a uniformly convergent sequence of functions without having to know the uniform limit in advance. To achieve this we need the following concept, which is the functional counterpart of a Cauchy sequence.

Definition 7.3. A sequence of functions $\{f_n\}$ on S is said to be **uniformly Cauchy** on S if, for every $\epsilon > 0$, there exists N such that

$$n, m \ge N \quad \text{implies} \quad |f_n(x) - f_m(x)| < \epsilon \quad \text{for all } x \text{ in } S.$$

The following **Cauchy criterion for uniform convergence** is the functional counterpart of Theorem 2.10.

Theorem 7.1. *A sequence of functions $\{f_n\}$ converges uniformly to a function f on S if and only if it is uniformly Cauchy on S.*

Ideas of Proof. Both directions are $\frac{\epsilon}{2}$-arguments similar to the proofs of Theorems 2.9 and 2.10.

Proof. For the "only if" part, suppose $\epsilon > 0$ is given. Then there exists N such that

$$n \geq N \quad \text{implies} \quad |f_n(x) - f(x)| < \frac{\epsilon}{2} \quad \text{for all } x \text{ in } S.$$

Thus, for $n, m \geq N$ and x in S, we have

$$|f_n(x) - f_m(x)| \leq |f_n(x) - f(x)| + |f(x) - f_m(x)| < \epsilon,$$

showing that $\{f_n\}$ is uniformly Cauchy.

To prove the "if" part, assume that $\{f_n\}$ is uniformly Cauchy on S. For the function f, first note that $\{f_n(x)\}$ is a Cauchy sequence, and hence a convergent sequence, for each x in S (Exercise (5) below). We may therefore define

$$f(x) = \lim_{n \to \infty} f_n(x) \quad \text{for each } x \text{ in } S.$$

To see that we have uniform convergence, suppose $\epsilon > 0$ is given. Then there exists N such that

$$n, m \geq N \quad \text{implies} \quad |f_n(x) - f_m(x)| < \frac{\epsilon}{2} \quad \text{for all } x \text{ in } S.$$

We have the inequality

$$|f_n(x) - f(x)| \leq |f_n(x) - f_m(x)| + |f_m(x) - f(x)|,$$

which holds for all m, n and x in S. The first term on the right is $< \frac{\epsilon}{2}$ for all x as long as $n, m \geq N$. Regardless of what x is, a large enough m, depending on x, will make the second term on the right also $< \frac{\epsilon}{2}$. Therefore, we have shown that $f_n \to f$ uniformly on S. $\square$

We will make use of Theorem 7.1 in the next section to show that uniform convergence preserves certain functional properties.

7.1.6 *Exercises*

(1) Write down explicitly what it means for a sequence of functions $\{f_n\}$ to not converge pointwise or uniformly to a function f on S.
(2) Prove that if $\{f_n\}$ converges uniformly to f on S, then f_n converges pointwise to f on S.

(3) Prove that $\{f_n\}$ converges uniformly to f on S if and only if

$$\lim_{n \to \infty} (\sup \{|f_n(x) - f(x)| : x \in S\}) = 0.$$

(4) Prove that the uniform limit of a uniformly convergent sequence of functions is unique. In other words, if $\{f_n\}$ converges uniformly to f on S and $\{f_n\}$ converges uniformly to g on S, then $f = g$ on S.

(5) Suppose $\{f_n\}$ is uniformly Cauchy on S. Prove that $\{f_n(x)\}$ is a Cauchy sequence for each x in S.

(6) In each case, determine whether $\{f_n\}$ converges pointwise, uniformly, or neither.

  (a) $f_n = \frac{x}{n}$ on $[0, b]$ or $[0, \infty)$.
  (b) $f_n = \frac{x}{1+nx}$ on $[0, 1]$ or $[0, 1)$.
  (c) $f_n = \frac{nx}{1+nx}$ on $[0, 1]$ or $[0, \infty)$.
  (d) $f_n = \frac{nx}{1+n^3x}$ on $[0, b]$ or $[0, \infty)$.
  (e) $f_n = \frac{x}{x+n}$ on $[0, \infty)$.
  (f) $f_n = \frac{nx}{e^{nx}}$ on $[0, b]$ or $[0, \infty)$.

(7) Prove that in Example 7.1 $f_n = x^n \to f = 0$ pointwise on $[0, 1)$.

(8) Prove that in Example 7.2 $\lim_{n \to \infty} f_n = \int_0^1 f$.

(9) In the context of Example 7.3:

  (a) Sketch the graph of f_n. Then prove that it is continuous on $[0, 1]$.
  (b) Prove that $\int_0^1 f_n = 1$.
  (c) Prove that $f_n \to f = 0$ pointwise on $[0, 1]$.

(10) In the context of Example 7.4:

  (a) Sketch the graph of f_n. Compare the graphs of f_1, f_2, and f_3.
  (b) Prove that $\lim_{n \to \infty} f_n'(x)$ does not exist for any x.
  (c) Prove that $f_n \to f = 0$ pointwise on $[0, 1]$.
  (d) Prove that $\lim_{n \to \infty} \int_0^1 f_n = \int_0^1 f$.

(11) In the context of Example 7.7:

  (a) Sketch the graph of f_n. Compare the graphs of f_1, f_2, and f_3.
  (b) Prove that $f_n \to f = 0$ uniformly on $[0, 1]$.
  (c) Prove that $\lim_{n \to \infty} \int_0^1 f_n = \int_0^1 f$.

(12) Suppose each f_n is an increasing function and that $\{f_n\}$ converges pointwise to f on $[a, b]$. Prove that f is increasing.

(13) Suppose $\{f_n\}$ converges uniformly to f on S with each f_n continuous and that $\{a_n\}$ is a sequence in S converging to a in S. Prove that the sequence $\{f_n(a_n)\}$ converges to $f(a)$.

(14) In the previous exercise, give an example to show that $\{f_n(a_n)\}$ may not converge to $f(a)$ if uniform convergence is replaced by pointwise convergence.

(15) Prove that a sequence $\{f_n\}$ is uniformly Cauchy if and only if, for every $\epsilon > 0$, there exists N such that

$$n > m \geq N \quad \text{implies} \quad |f_n(x) - f_m(x)| < \epsilon \quad \text{for all} \quad x \in S.$$

7.2   Interchange of Limits

In this section we discuss several functional properties that are preserved by uniform convergence. As explained in the previous section, each such preservation result can be interpreted as a commutation property between the limit operation $\lim_{n\to\infty}$ and the functional property under consideration. Another main theme in this section is that we will recycle essentially the same $\frac{\epsilon}{3}$-argument many times.

7.2.1   *Preservation of Limits*

Example 7.1 shows that pointwise convergence is not strong enough to force the limit operation $\lim_{x\to c}$ to pass from f_n to the pointwise limit. The following result shows that uniform convergence is strong enough for this purpose.

Theorem 7.2. *Suppose f_n converges uniformly to f on S, c is a limit point of S, and that the limit $a_n = \lim_{x\to c} f_n(x)$ exists for each n. Then the sequence $\{a_n\}$ is convergent and*

$$\lim_{x\to c} f(x) = \lim_{n\to\infty} a_n.$$

Ideas of Proof. Both assertions are $\frac{\epsilon}{3}$-arguments. To see that $\{a_n\}$ is Cauchy, we use the estimates

$$a_n \approx f_n(x) \approx f_m(x) \approx a_m.$$

Here each wavy equal sign $\approx$ means that the two quantities next to it can be closely approximated. The first and the last $\approx$ hold if x is sufficiently close to c, while the middle $\approx$ holds for large enough n and m. For the equality of limits, we use the estimates

$$\lim_{n\to\infty} a_n \approx a_n \approx f_n(x) \approx f(x).$$

The first and the last $\approx$ hold for large enough n, while the middle $\approx$ holds when x is close enough to c.

Proof.   To prove that $\{a_n\}$ is convergent, it suffices to show that it is a Cauchy sequence. Given $\epsilon > 0$, the uniform convergence assumption and Theorem 7.1 imply that there exists N such that

$$n, m \ge N \quad \text{implies} \quad |f_n(x) - f_m(x)| < \frac{\epsilon}{3} \quad \text{for all } x \text{ in } S.$$

For any such n and m, we also know that there exist δ_1 and $\delta_2 > 0$ such that

$$0 < |x - c| < \delta_1 \quad \text{implies} \quad |f_n(x) - a_n| < \frac{\epsilon}{3}$$

and

$$0 < |x - c| < \delta_2 \quad \text{implies} \quad |f_m(x) - a_m| < \frac{\epsilon}{3},$$

provided x is in S. Thus, as long as $n, m \geq N$ and $0 < |x - c| < \min\{\delta_1, \delta_2\}$ with x in S, the inequality

$$|a_n - a_m| \leq |a_n - f_n(x)| + |f_n(x) - f_m(x)| + |f_m(x) - a_m|$$

implies

$$|a_n - a_m| < \epsilon.$$

Therefore, $\{a_n\}$ is a Cauchy sequence, and hence a convergent sequence. The assertion about the equality of limits is proved by a similar $\frac{\epsilon}{3}$-argument, which is left as an exercise. $\square$

The equality of limits in Theorem 7.2 can be written as

$$\lim_{x \to c} \lim_{n \to \infty} f_n(x) = \lim_{n \to \infty} \lim_{x \to c} f_n(x).$$

In other words, under the hypothesis of uniform convergence, the limit operations $\lim_{n \to \infty}$ and $\lim_{x \to c}$ commute.

7.2.2   Preservation of Integrals

Example 7.3 shows that pointwise convergence is not strong enough to force the integral to commute with the limit operation $\lim_{n \to \infty}$. The following result shows that uniform convergence is strong enough for this purpose.

Theorem 7.3. *Suppose f_n converges uniformly to f on $[a, b]$, and each f_n is integrable on $[a, b]$. Then f is also integrable on $[a, b]$, and*

$$\int_a^b f = \lim_{n \to \infty} \int_a^b f_n.$$

Ideas of Proof. The integrability of f is again proved by an $\frac{\epsilon}{3}$-argument, using the approximations

$$U(f; P) \approx U(f_n; P) \approx L(f_n; P) \approx L(f; P).$$

The two outer $\approx$ hold for large enough n, while the middle $\approx$ holds for partitions P with small enough norms. The equality is proved using the approximation $f \approx f_n$ for large enough n and Corollary 6.3.

Proof.   To prove that f is integrable on $[a, b]$, suppose $\epsilon > 0$ is given. By uniform convergence there exists N such that

$$n \geq N \quad \text{implies} \quad |f_n(x) - f(x)| < \frac{\epsilon}{3(b-a)} \quad \text{for all } x \text{ in } [a, b].$$

Fix such an n, for example, $n = N$. The integrability of f_n implies the existence of a partition P of $[a, b]$ such that

$$U(f_n; P) - L(f_n; P) < \frac{\epsilon}{3}.$$

In each sub-interval of P, we have

$$|u_i(f) - u_i(f_n)| \le \frac{\epsilon}{3(b-a)} \quad \text{and} \quad |l_i(f) - l_i(f_n)| \le \frac{\epsilon}{3(b-a)}.$$

Therefore, the inequalities

$$U(f;P) - L(f;P)$$
$$\le |U(f;P) - U(f_n;P)| + |U(f_n;P) - L(f_n;P)| + |L(f_n;P) - L(f;P)|$$
$$< \sum |u_i(f) - u_i(f_n)|\Delta x_i + \frac{\epsilon}{3} + \sum |l_i(f_n) - l_i(f)|\Delta x_i$$
$$\le \frac{\epsilon}{3(b-a)}(b-a) + \frac{\epsilon}{3} + \frac{\epsilon}{3(b-a)}(b-a) = \epsilon$$

imply that f is integrable on $[a,b]$.

For the equality, suppose $\epsilon > 0$ is given. Then uniform convergence implies that there exists N such that

$$n \ge N \quad \text{implies} \quad |f_n(x) - f(x)| < \frac{\epsilon}{b-a} \quad \text{for all } x \text{ in } [a,b].$$

For such n, the inequalities

$$-\frac{\epsilon}{b-a} + f_n(x) < f(x) < \frac{\epsilon}{b-a} + f_n(x)$$

and Corollary 6.3 imply

$$-\epsilon + \int_a^b f_n \le \int_a^b f \le \epsilon + \int_a^b f_n.$$

This is equivalent to

$$\left| \int_a^b f - \int_a^b f_n \right| \le \epsilon.$$

The desired equality follows. $\qquad\square$

The equality in Theorem 7.3 can be written as

$$\int_a^b \lim_{n\to\infty} f_n = \lim_{n\to\infty} \int_a^b f_n.$$

In other words, under the hypothesis of uniform convergence, the integral $\int_a^b$ and the limit operation $\lim_{n\to\infty}$ commute.

7.2.3 Preservation of Continuity

Example 7.2 shows that pointwise convergence is not strong enough to preserve continuity. The following result shows that uniform convergence is strong enough for this purpose.

Theorem 7.4. *Suppose f_n converges uniformly to f on S, and each f_n is continuous at a point c in S. Then f is also continuous at c.*

Ideas of Proof. It is possible to obtain this result using Theorem 7.2; the details are left as an exercise. Here we provide a direct proof using an $\frac{\epsilon}{3}$-argument via the approximations

$$f(x) \approx f_n(x) \approx f_n(c) \approx f(c).$$

The two outer $\approx$ hold for large enough n. The middle $\approx$ holds for x close to c.

Proof.   Given $\epsilon > 0$, uniform convergence implies there exists N such that

$$n \geq N \quad \text{implies} \quad |f(x) - f_n(x)| < \frac{\epsilon}{3} \quad \text{for all } x \text{ in } S.$$

Fix such an n, for example, $n = N$. Continuity of f_n at c implies there exists $\delta > 0$ such that

$$|x - c| < \delta \quad \text{implies} \quad |f_n(x) - f_n(c)| < \frac{\epsilon}{3},$$

provided x is in S. Therefore, the inequalities

$$|f(x) - f(c)| \leq |f(x) - f_n(x)| + |f_n(x) - f_n(c)| + |f_n(c) - f(c)| < \epsilon$$

hold whenever $|x - c| < \delta$ with x in S. This shows that f is continuous at c.   □

When c is a limit point of S, using (4.1) the conclusion of the previous theorem can be written as

$$\lim_{x \to c} \lim_{n \to \infty} f_n(x) = \lim_{x \to c} f(x) = f(c) = \lim_{n \to \infty} f_n(c) = \lim_{n \to \infty} \lim_{x \to c} f_n(x).$$

In other words, under the hypothesis of uniform convergence, the limit operations $\lim_{n \to \infty}$ and $\lim_{x \to c}$ commute.

7.2.4   *Preservation of Derivatives*

Examples 7.4 and 7.6 together show that uniform convergence is not strong enough for the preservation of derivatives. One way to ensure derivatives are preserved is to have uniform convergence for the derivatives and convergence at a single point.

Theorem 7.5. *Suppose each f_n is differentiable on an open interval I containing $[a, b]$, there exists a point α in $[a, b]$ such that $\{f_n(\alpha)\}$ is a convergent sequence, and f_n' converges uniformly to a function g on $[a, b]$. Then:*

(1) f_n converges uniformly to a function f on $[a, b]$.
(2) f is differentiable on (a, b) with $f' = g$.

Ideas of Proof.   The Mean Value Theorem 5.7 is a way to relate a function and its derivative, so it is no surprise that we will use it. The first assertion is proved by an $\frac{\epsilon}{2}$-argument. The other assertion is proved by an $\frac{\epsilon}{3}$-argument via the approximations

$$\frac{f(x) - f(d)}{x - d} \approx \frac{f_m(x) - f_m(d)}{x - d} \approx f_m'(d) \approx g(d).$$

The last $\approx$ holds for large enough m, while the middle $\approx$ holds for x close enough to d. The first $\approx$ requires a bit of work.

Proof. For the first assertion, we will show that $\{f_n\}$ is uniformly Cauchy, which is enough by Theorem 7.1. For each point x in $[a, b]$ and any two indices n and m, the Mean Value Theorem applied to $f_n - f_m$ yields a point c between x and α such that

$$f_n(x) - f_m(x) = [f_n(\alpha) - f_m(\alpha)] + (x - \alpha)[f'_n(c) - f'_m(c)].$$

We know that $\{f_n(\alpha)\}$ is a Cauchy sequence, and $\{f'_n\}$ is uniformly Cauchy. Thus, given $\epsilon > 0$, there exist N_1 and N_2 such that

$$n, m \geq N_1 \quad \text{implies} \quad |f_n(\alpha) - f_m(\alpha)| < \frac{\epsilon}{2}$$

and

$$n, m \geq N_2 \quad \text{implies} \quad |f'_n(y) - f'_m(y)| < \frac{\epsilon}{2(b - a)} \quad \text{for all } y \text{ in } S.$$

Since $|x - \alpha| \leq b - a$, for $n, m \geq \max\{N_1, N_2\}$, we have

$$|f_n(x) - f_m(x)| \leq |f_n(\alpha) - f_m(\alpha)| + |x - \alpha||f'_n(c) - f'_m(c)|$$
$$< \frac{\epsilon}{2} + (b - a)\frac{\epsilon}{2(b - a)} = \epsilon.$$

This proves that $\{f_n\}$ converges uniformly to a function f on S.

Pick a point d in (a, b), and we will show that $f'(d) = g(d)$. For any two indices n and m, the Mean Value Theorem applied to $f_n - f_m$ on the interval determined by d and $x \neq d$ in (a, b) yields a point e between d and x such that

$$\frac{f_n(x) - f_n(d)}{x - d} - \frac{f_m(x) - f_m(d)}{x - d} = f'_n(e) - f'_m(e).$$

Given $\epsilon > 0$, using the above equality, the uniform convergence of $\{f'_k\}$ now implies there exists N_1 such that

$$n, m \geq N_1 \quad \text{implies} \quad \left| \frac{f_n(x) - f_n(d)}{x - d} - \frac{f_m(x) - f_m(d)}{x - d} \right| < \frac{\epsilon}{3}.$$

At this moment d and x are fixed. So keeping $m \geq N_1$ fixed and taking the limit $\lim_{n \to \infty}$ in the above inequality, we obtain

$$\left| \frac{f(x) - f(d)}{x - d} - \frac{f_m(x) - f_m(d)}{x - d} \right| \leq \frac{\epsilon}{3}.$$

Since $g(d) = \lim_{m \to \infty} f'_m(d)$, there exists N_2 such that

$$m \geq N_2 \quad \text{implies} \quad |f'_m(d) - g(d)| < \frac{\epsilon}{3}.$$

Now set $m = \max\{N_1, N_2\}$, so the previous two inequalities both hold. The differentiability of f_m at d implies the existence of $\delta > 0$ such that

$$0 < |x - d| < \delta \quad \text{implies} \quad \left| \frac{f_m(x) - f_m(d)}{x - d} - f'_m(d) \right| < \frac{\epsilon}{3},$$

provided x is in (a, b). Therefore, for such x, the last three $\frac{\epsilon}{3}$-inequalities together imply

$$\left| \frac{f(x) - f(d)}{x - d} - g(d) \right| < \epsilon.$$

This proves the differentiability of f at d and $f'(d) = g(d)$. $\qquad\qquad\square$

The assertion $f' = g$ in the previous theorem can be written as

$$\frac{d}{dx}\left(\lim_{n\to\infty} f_n \right) = \lim_{n\to\infty} \frac{df_n}{dx}.$$

In other words, under the stated hypotheses, the differential operator $\frac{d}{dx}$ and the limit $\lim_{n\to\infty}$ commute.

7.2.5 Exercises

(1) If $\{f_n\}$ and $\{g_n\}$ converge uniformly on S to f and g, respectively, prove that $\{f_n + g_n\}$ converges uniformly on S to $f + g$.
(2) In the previous exercise, give an example to show that $\{f_n g_n\}$ may not converge uniformly on S to fg.
(3) Suppose $\{f_n\}$ and $\{g_n\}$ converge uniformly on S to f and g, respectively, and that all the f_n and g_n are bounded on S. Prove that $\{f_n g_n\}$ converges uniformly on S to fg.
(4) Give an example with the following properties. The sequence $\{f_n\}$ converges to f pointwise on an interval $[a, b]$, each f_n is integrable on $[a, b]$, and f is not integrable on $[a, b]$.
(5) Suppose f_n converges uniformly to f on S, and each f_n is uniformly continuous on S. Prove that f is uniformly continuous on S.
(6) Prove the equality of limits in Theorem 7.2. Also, where exactly in that proof is the hypothesis that c is a limit point of S used?
(7) Prove the following statement that was used in the proof of Theorem 7.3. If

$$|g(x) - f(x)| < \delta \quad \text{for all} \quad x \in S,$$

then

$$|\sup\{g(x) : x \in S\} - \sup\{f(x) : x \in S\}| \le \delta$$

and

$$|\inf\{g(x) : x \in S\} - \inf\{f(x) : x \in S\}| \le \delta.$$

(8) The proof of Theorem 7.4 uses the ϵ-δ characterization of continuity. Write down a version of that proof that uses Definition 4.5 of continuity.
(9) Use Theorem 7.2 to give another proof of Theorem 7.4.
(10) Give an example to show that the conclusion in Theorem 7.2 may still hold even if f_n does not converge uniformly to f.

(11) Give an example to show that the conclusion in Theorem 7.3 may still hold even if f_n does not converge uniformly to f.

(12) Give an example in which f_n is not continuous at any point on $[0,1]$, and f_n converges uniformly to a continuous function on $[0,1]$.

(13) Suppose $\{f_n\}$ converges uniformly to f on S and that each f_n is bounded on S. Prove that f is bounded on S.

(14) In the previous exercise, give an example to show that f may not be bounded if uniform convergence is replaced by pointwise convergence.

(15) A sequence of functions $\{f_n\}$ on S is called **uniformly bounded** if there exists a real number M such that $|f_n(x)| \le M$ for all n and x in S. Suppose $\{f_n\}$ converges uniformly to f on S and that each f_n is bounded on S. Prove that $\{f_n\}$ is uniformly bounded.

(16) In the previous exercise, give an example to show that $\{f_n\}$ may not be uniformly bounded if uniform convergence is replaced by pointwise convergence.

7.3   Series of Functions

As discussed in chapter 2.5, a series is a particular kind of a sequence in which the nth term is a partial sum of n quantities. In this section we consider series of functions. We first establish a Cauchy Criterion, the Weierstrass M-Test, and some results about interchange of limits for series of functions. Then we construct a function that is continuous on $\mathbb{R}$ but is nowhere differentiable.

7.3.1   *Definition*

Definition 7.4. Suppose $\{f_n\}$ is a sequence of functions on S. The sequence of functions $\{s_n\}$ on S with each

$$s_m = f_1 + \cdots + f_m = \sum_{i=1}^{m} f_i$$

is called a **series of functions**, also written as $\sum f_n$. The function s_m is called the m**th partial sum**. If $\{s_n\}$ converges pointwise or uniformly to a function f on S, then we also write

$$f = \sum f_n.$$

Convergence results about series of real numbers and sequences of functions can now be used to obtain convergence results for series of functions.

7.3.2   *Cauchy Criterion*

Applying Theorem 7.1 to a series of functions, we obtain the following Cauchy criterion for uniform convergence.

Corollary 7.1. *A series of functions $\sum f_n$ converges uniformly to a function on S if and only if, for every $\epsilon > 0$, there exists N such that*

$$n > m \geq N \quad implies \quad \left| \sum_{i=m+1}^{n} f_i(x) \right| < \epsilon \quad for \ all \ x \ in \ S.$$

Proof.   First observe that

$$s_n - s_m = \sum_{i=m+1}^{n} f_i$$

for $n > m$. Therefore, the stated condition says exactly that the sequence of functions $\{s_n\}$ is uniformly Cauchy, which by Theorem 7.1 is equivalent to uniform convergence.   □

A particularly useful consequence of the previous Cauchy criterion is the following test for uniform convergence.

7.3.3   Weierstrass M-Test

Corollary 7.2. *Suppose for each n there exists a real number M_n such that*

$$|f_n(x)| \leq M_n \quad for \ all \ x \ in \ S.$$

If the series $\sum M_n$ converges, then the series of functions $\sum f_n$ converges uniformly on S.

Proof.   By Theorem 3.1, for every $\epsilon > 0$, there exists N such that $n > m \geq N$ implies

$$\left| \sum_{i=m+1}^{n} f_i(x) \right| \leq \sum_{i=m+1}^{n} |f_i(x)| \leq \sum_{i=m+1}^{n} M_i < \epsilon$$

for all x in S. Corollary 7.1 now says that the series of functions $\sum f_n$ converges uniformly.   □

Example 7.8. By the Weierstrass M-Test, the series

$$\sum \frac{1}{(x+n)^p}$$

converges uniformly on $[0, \infty)$ if $p > 1$, since $\sum \frac{1}{n^p}$ is a convergent p-series.

Be careful that, unlike the Cauchy criterion, the Weierstrass M-Test is a one-way implication. So if the series $\sum M_n$ diverges, then there is no conclusion to be made from the Weierstrass M-Test.

7.3.4 *Interchange of Limits*

Results about interchange of limits in the previous section can be applied to series of functions. The proofs of the next few results are immediate applications of the corresponding results in the previous section.

   The following result is the special case of Theorem 7.2 when applied to a series of functions.

Corollary 7.3. *Suppose a series of functions $\sum f_n$ converges uniformly to f on S, c is a limit point of S, and that the limit $a_n = \lim_{x \to c} f_n(x)$ exists for each n. Then the series $\sum a_n$ is convergent and*

$$\lim_{x \to c} f(x) = \sum a_n.$$

   The equality of the previous corollary may be written as

$$\lim_{x \to c} \left(\sum f_n(x) \right) = \sum \left(\lim_{x \to c} f_n(x) \right).$$

In other words, under the hypothesis of uniform convergence, the limit operation $\lim_{x \to c}$ and the series operation $\sum$ commute. The next result is the special case of Theorem 7.3 when applied to a series of functions.

Corollary 7.4. *Suppose a series of functions $\sum f_n$ converges uniformly to f on $[a,b]$, and each f_n is integrable on $[a,b]$. Then f is also integrable on $[a,b]$, the series $\sum \left(\int_a^b f_n \right)$ is convergent, and*

$$\int_a^b f = \sum \left(\int_a^b f_n \right).$$

   The equality of the previous corollary may be written as

$$\int_a^b \left(\sum f_n \right) = \sum \left(\int_a^b f_n \right).$$

In other words, under the hypothesis of uniform convergence, the integral $\int_a^b$ and the series operation $\sum$ commute. This is saying that the integral of the limit can be computed by first integrating each term. The following result is the special case of Theorem 7.4 when applied to a series of functions.

Corollary 7.5. *Suppose a series of functions $\sum f_n$ converges uniformly to f on S, and each f_n is continuous at a point c in S. Then f is also continuous at c.*

   When c is a limit point of S, the conclusion of the previous theorem can be written as

$$\lim_{x \to c} \left(\sum f_n(x) \right) = \sum \left(\lim_{x \to c} f_n(x) \right).$$

The next result is the special case of Theorem 7.5 when applied to a series of functions.

Corollary 7.6. *Suppose each f_n is differentiable on an open interval I containing $[a, b]$, there exists a point α in $[a, b]$ such that $\sum f_n(\alpha)$ is a convergent series, and $\sum f_n'$ converges uniformly to a function g on $[a, b]$. Then:*

(1) $\sum f_n$ converges uniformly to a function f on $[a, b]$.
(2) f is differentiable on (a, b) with $f' = \sum f_n'$.

The equality in the previous corollary may be written as

$$\frac{d}{dx}\left(\sum f_n\right) = \sum\left(\frac{df_n}{dx}\right).$$

In other words, under the stated hypotheses, the differential operator $\frac{d}{dx}$ commutes with the series operation $\sum$. This is saying that the derivative of the limit can be computed by differentiating each term f_n.

7.3.5   A Continuous Nowhere Differentiable Functions

Here we construct a continuous function on $\mathbb{R}$ that is not differentiable at any point. There are two purposes of this example. First, it shows that some weird things can happen for a continuous function. In fact, it almost defies the imagination that a continuous function on $\mathbb{R}$ can be nowhere differentiable. Second, the construction is a good illustration of the Weierstrass M-Test and Corollary 7.5.

Let us first explain the ideas for the construction. We will construct a continuous function $f = \sum f_n$. With carefully chosen values h_n that converge to 0, the difference quotients $\frac{1}{h_n}(f(x + h_n) - f(x))$ can be made to jump between odd and even integers as n increases. This suffices to show that f is not differentiable at x. For the actual construction, we need the following definition.

Definition 7.5. Let p be a positive real number. A function f on $\mathbb{R}$ is called **p-periodic** if it satisfies

$$f(x + p) = f(x)$$

for all x in $\mathbb{R}$.

A p-periodic function is uniquely determined by its values in the interval $[0, p]$. Conversely, given a function f on $[0, p]$ with $f(0) = f(p)$, there exists a unique p-periodic function on $\mathbb{R}$ that agrees with f on $[0, p]$.

Now let f_0 be the 2-periodic function uniquely determined by

$$f_0(x) = \begin{cases} x & \text{if } 0 \leq x \leq 1, \\ 2 - x & \text{if } 1 \leq x \leq 2. \end{cases}$$

For $k \geq 1$ define

$$f_k(x) = \frac{1}{4^k} f_0(4^k x),$$

which is $\frac{2}{4^k}$-periodic. Graphically, f_0 consists of the $[-1,1]$ portion of the graph of $|x|$, which is then made 2-periodic. Over $[0,2]$ its graph has the shape of a pyramid of height 1. It consists of two lines, one with slope 1 and the other with slope -1. Over $[0, \frac{1}{2}]$ the graph of f_1 is a pyramid of height $\frac{1}{4}$, and four of these are inscribed inside the pyramid of f_0 over $[0,2]$. The lines in its graph have slopes either 1 or -1. In general, the graph of f_{k+1} consists of little pyramids that are four times as small, both in width and in height, as those in f_k.

Each function f_k is continuous on $\mathbb{R}$ and satisfies

$$|f_k(x)| \le \frac{1}{4^k} \quad \text{for all } x \text{ in } \mathbb{R}.$$

Therefore, the Weierstrass M-Test (Corollary 7.2) says the series of functions $\sum_{i=0}^{\infty} f_i$ converges uniformly to a function f on $\mathbb{R}$. Moreover, by Corollary 7.5 the function f is continuous on $\mathbb{R}$.

Now we observe that f is not differentiable at any point. Pick a point c in $\mathbb{R}$, and we will show that f is not differentiable at c. For each integer $k \ge 0$, define

$$h_k = \pm \frac{2}{4^{k+1}},$$

where the sign $\pm$ is chosen such that both $4^k c$ and $4^k(c + h_k)$ lie in the interval $[m, m+1]$ for some integer m. This is possible because $4^k h_k = \pm \frac{1}{2}$. For each k, the choice of h_k and the $\frac{2}{4^i}$-periodicity of f_i imply

$$f_i(c + h_k) - f_i(c) = \begin{cases} \pm h_k & \text{if } 0 \le i \le k, \\ 0 & \text{if } i > k. \end{cases} \tag{7.2}$$

For example, we have

$$f_{k+1}(c + h_k) - f_{k+1}(c) = 0$$

because f_{k+1} is $\frac{2}{4^{k+1}}$-periodic. Likewise, we have

$$f_k(c + h_k) - f_k(c_k) = \frac{f_0(4^k(c + h_k)) - f_0(4^k c)}{4^k} = \pm h_k.$$

Therefore, the relevant difference quotient of f is

$$\frac{f(c + h_k) - f(c)}{h_k} = \sum_{i=0}^{k} \frac{f_i(c + h_k) - f_i(c)}{h_k} = \sum_{i=0}^{k} \pm 1. \tag{7.3}$$

This last quantity is odd when k is even, and it is even when k is odd. Therefore, this sequence of difference quotients does not converge. Since the sequence $\{h_k\}$ converges to 0, we conclude that $\lim_{h \to 0} \frac{f(c+h) - f(c)}{h}$ does not exist.

7.3.6 *Exercises*

(1) Determine whether the following series converge uniformly.

   (a) $\sum \frac{\cos n^2 x}{n^2}$ on $\mathbb{R}$.

(b) $\sum \frac{1}{x^n}$ on $[1,2]$ or $[2,\infty)$.

(c) $\sum \frac{n}{x^n}$ on $[1,2]$ or $[2,\infty)$.

(d) $\sum x^n$ on $[-r,r]$ for $0 < r < 1$ or $(-1,1)$.

(e) $\sum \frac{x^n}{n^2}$ on $[-1,1]$ or $\mathbb{R}$.

(f) $\sum \left(\frac{x}{2}\right)^n$ on $[-r,r]$ for $0 < r < 2$ or $(-2,2)$.

(2) Write down the details of the proof of Corollary 7.3.

(3) Write down the details of the proof of Corollary 7.4.

(4) Write down the details of the proof of Corollary 7.5.

(5) Write down the details of the proof of Corollary 7.6.

(6) Suppose $\sum f_n$ converges uniformly to f on S, and each f_n is uniformly continuous on S. Prove that f is also uniformly continuous on S.

(7) Suppose $\sum f_n$ and $\sum g_n$ converge uniformly to f and g on S, respectively. Prove that $\sum(f_n + g_n)$ converges uniformly to $f + g$ on S.

(8) Suppose $\sum f_n$ converges uniformly to f on S. Prove that the sequence of functions $\{f_n\}$ converges uniformly to the 0 function. This is the functional analog of Corollary 3.2.

(9) Prove that a p-periodic function f is uniquely determined by its values in $[0,p]$, and $f(0) = f(p)$.

(10) Suppose f is a function on $[0,p]$ such that $f(0) = f(p)$. Prove that there exists a unique p-periodic function g on $\mathbb{R}$ such that $g(x) = f(x)$ for x in $[0,p]$. The function g is called the **periodic extension** of f.

(11) Suppose f is a p-periodic function, and k is a positive integer. Prove that f is kp-periodic.

(12) Suppose f is a p-periodic function, and r is a positive real number. Prove that $g(x) = f(rx)$ is $\frac{p}{r}$-periodic.

(13) In section 7.3.5:

(a) Sketch the graphs of f_0, f_1, and f_2.

(b) Prove that each f_k is continuous on $\mathbb{R}$.

(c) For each point c in $\mathbb{R}$ and integer $k \geq 0$, prove that there exists a sign, + or $-$, such that both $4^k c$ and $4^k(c \pm \frac{2}{4^{k+1}})$ lie in the interval $[m, m+1]$ for some integer m.

(d) Justify (7.2) and (7.3).

(14) This exercise provides another continuous nowhere differentiable function using a variation of the construction in section 7.3.5. With the same notations there, define

$$g(x) = \sum_{k=0}^{\infty} \left(\frac{3}{4}\right)^k f_0(4^k x).$$

(a) Prove that g is continuous on $\mathbb{R}$.

(b) Prove that

$$\left| \frac{g(c + h_k) - g(c)}{h_k} \right| \geq 3^k - \sum_{i=0}^{k-1} 3^i = \frac{3^k + 1}{2}.$$

(c) Use the previous part to conclude that g is not differentiable at c.

7.4 Power Series

In this section, we discuss power series. They are particular examples of series of functions that have very nice properties. We will study two issues. First, given a power series, find where it is pointwise or uniformly convergent, and see whether its derivative and integral can be computed term-wise. Second, given a function, find where it can be represented by a power series. Let us begin with some definitions.

7.4.1 *Definitions*

Definition 7.6. Given a point c in $\mathbb{R}$, a **power series at** c is a series of functions of the form

$$\sum_{i=0}^{\infty} a_i (x - c)^i = a_0 + a_1 (x - c) + \cdots,$$

where each a_i is a real number.

We will often omit "at c" when we talk about power series. Observe that a power series is a series of functions $\sum f_i$ in which $f_i = a_i (x - c)^i$ is a degree i polynomial, provided $a_i \neq 0$. In particular, each f_i is defined on $\mathbb{R}$ and has all higher derivatives. Moreover, a power series at c must converge at the point $x = c$, since $f_i(c) = 0$ for all $i > 0$. The main question is where a power series converges besides the point $x = c$. First recall from section 3.4 the concept of absolute convergence, the Root Test, and the Ratio Test. We will also use the concept of limit superior discussed in section 2.4.

Let us briefly motivate the next definition. Since $f_i = a_i (x - c)^i$ is a degree i polynomial, this suggests a comparison with a geometric series $\sum r^i$, so we have the estimate

$$a_i (x - c)^i \approx r^i.$$

The geometric series is convergent exactly when $|r| < 1$. Thus, taking the ith root above suggests that a condition similar to

$$|x - c| < \frac{1}{|a_i|^{\frac{1}{i}}} \quad \text{for large } i$$

should guarantee the absolute convergence of the power series. This leads to the following definitions.

Definition 7.7. For a power series $\sum a_i (x - c)^i$, write $L = \limsup |a_n|^{\frac{1}{n}}$. Define its **radius of convergence** as the extended real number

$$R = \begin{cases} 0 & \text{if } L = \infty, \\ \frac{1}{L} & \text{if } 0 < L < \infty, \\ +\infty & \text{if } L = 0. \end{cases}$$

Its **interval of convergence** is defined as the open interval $(c - R, c + R)$ if $R > 0$, and as the single point $\{c\}$ if $R = 0$.

7.4.2   *Convergence*

The discussion just before the previous definition is made precise in the following convergence theorem.

Theorem 7.6. *Suppose $\sum a_i(x-c)^i$ is a power series with radius of convergence R.*

(1) The power series is absolutely convergent for x in the interval of convergence.
(2) The power series is divergent if $|x - c| > R$.

Proof.   We will prove the case
$$0 < L = \limsup |a_n|^{\frac{1}{n}} < \infty.$$
The cases $L = 0$ and $L = \infty$ are exercises. By the Root Test (Theorem 3.7), the power series is absolutely convergent if
$$\limsup |a_n(x - c)^n|^{\frac{1}{n}} = |x - c| \left(\limsup |a_n|^{\frac{1}{n}} \right) = |x - c| L < 1.$$
This is equivalent to saying x is in the interval of convergence. The Root Test also says that the power series is divergent if
$$\limsup |a_n(x - c)^n|^{\frac{1}{n}} = |x - c| L > 1.$$
This proves the theorem.   □

As the reader learned in calculus, the convergence behavior of a power series at the end points $c \pm R$ of the interval of convergence needs to be determined separately. The convergence tests developed in chapter 2.5 can be used for this purpose. One must be careful that Theorem 7.6 only tells us where the power series converges *pointwise*, namely, the interval of convergence, possibly including the end points. It does not say where it converges uniformly, which we will discuss shortly.

To determine the radius and interval of convergence, one has to compute $\limsup |a_n|^{\frac{1}{n}}$, which may not be an easy task. In practice one can often get away with an easier computation as follows. As a consequence of Theorem 2.14 and Exercise (6) on page 76, if the limit
$$L' = \lim_{n \to \infty} \left| \frac{a_{n+1}}{a_n} \right| \tag{7.4}$$
exists as an extended real number, then it is equal to $\lim |a_n|^{\frac{1}{n}}$. Therefore, we have the following result about the radius of convergence.

Theorem 7.7. *Suppose $\sum a_i(x-c)^i$ is a power series such that the limit L' in (7.4) exists as an extended real number. Then the radius of convergence is equal to*
$$\begin{cases} 0 & \text{if } L' = \infty, \\ \frac{1}{L'} & \text{if } 0 < L' < \infty, \\ +\infty & \text{if } L' = 0. \end{cases}$$

7.4.3   *Uniform Convergence*

In section 7.3.4 we discussed several properties that are preserved by uniform convergence. The following observation will allow us to use those results.

Theorem 7.8. *A power series $\sum a_i(x - c)^i$ converges uniformly on every closed bounded interval inside its interval of convergence.*

Ideas of Proof. We will compare the power series to a convergent geometric series and then use the Weierstrass M-Test to conclude its uniform convergence.

Proof. Once again we leave the $R = 0$ and $R = +\infty$ cases as exercises. We will prove the case $0 < R < +\infty$. A closed bounded interval inside $(c - R, c + R)$ must be contained in a closed bounded interval

$$I = [c - \epsilon R, c + \epsilon R] \quad \text{for some } 0 < \epsilon < 1.$$

Pick any r with $\epsilon < r < 1$. For x in I, we have

$$|x - c| < rR = \frac{r}{\limsup |a_n|^{\frac{1}{n}}},$$

which implies

$$\limsup \left(|a_n|^{\frac{1}{n}} |x - c| \right) < r.$$

Therefore, there exists N such that, for x in I, we have

$$|a_n(x - c)^n| < r^n \quad \text{for all } n \geq N.$$

Since $|r| < 1$, the geometric series $\sum r^n$ is convergent. The Weierstrass M-Test (Corollary 7.2) now implies that the power series converges uniform on I.   □

In particular, when the radius of convergence is $+\infty$, the power series converges uniformly on every closed bounded interval in $\mathbb{R}$. Next we turn to properties of power series that can be computed term-wise.

7.4.4   *Term-Wise Integration*

The following result says that a power series can be integrated term-by-term within its interval of convergence.

Theorem 7.9. *Suppose a power series $f(x) = \sum a_i(x-c)^i$ has radius of convergence $R > 0$. Then*

$$\int_c^x f(t)dt = \sum_{i=0}^{\infty} \frac{a_i}{i + 1}(x - c)^{i+1} \quad \text{for all } |x| < R,$$

and this power series also has radius of convergence R.

Proof. The equality follows from Theorem 6.9, Corollary 7.4, and Theorem 7.8, applied to the closed bounded interval with end points c and x. The assertion concerning the radius of convergence follows from the equality

$$\limsup |a_n|^{\frac{1}{n}} = \limsup \left| \frac{a_{n-1}}{n} \right|^{\frac{1}{n}},$$

which in turn is true because $\lim n^{\frac{1}{n}} = 1$ (Exercise (13) on page 44).   □

7.4.5   Term-Wise Differentiation

The following result says that a power series can be differentiated term-by-term within its interval of convergence.

Theorem 7.10. *Suppose a power series $f(x) = \sum a_i(x - c)^i$ has radius of convergence $R > 0$. Then:*

(1) The power series f is differentiable in its interval of convergence.
(2) Its derivative is given by

$$f'(x) = \sum_{i=1}^{\infty} i a_i (x - c)^{i-1}$$

for x in the interval of convergence $(c - R, c + R)$.
(3) The power series $\sum i a_i(x - c)^{i-1}$ also has radius of convergence R.

Proof. We apply Theorem 7.8 and Corollary 7.6 with $f_n = a_n(x-c)^n$. To see that we can actually use Corollary 7.6, it suffices to show that the term-wise differentiated power series $\sum i a_i(x - c)^{i-1}$ has radius of convergence R. To do this, it is enough to show

$$\limsup |a_n|^{\frac{1}{n}} = \limsup |(n + 1)a_{n+1}|^{\frac{1}{n}}.$$

This equality holds because $\lim(n + 1)^{\frac{1}{n}} = 1$.   □

Since the term-wise differentiated power series has the same radius of convergence, the same process can be applied to it to obtain the second derivative, and so forth.

Corollary 7.7. *Suppose a power series $f(x) = \sum a_i(x-c)^i$ has radius of convergence $R > 0$. Then:*

(1) The nth derivative $f^{(n)}$ exists for each $n \geq 1$ in its interval of convergence.
(2) Its nth derivative is given by

$$f^{(n)}(x) = \sum_{i=n}^{\infty} \frac{i!}{(i - n)!} a_i (x - c)^{i-n}$$

for x in the interval of convergence $(c - R, c + R)$.
(3) The power series in the previous part also has radius of convergence R.

Proof. The case $n = 1$ is Theorem 7.10. The induction step also follows from Theorem 7.10.   □

Therefore, a power series is infinitely differentiable in its interval of convergence, and the derivatives are all computed term-wise. Recall that a power series at c must converge at $x = c$. The following result was informally suggested in section 5.3.1.

Corollary 7.8. *Suppose a power series $f(x) = \sum a_i(x-c)^i$ has radius of convergence $R > 0$. Then*

$$a_n = \frac{f^{(n)}(c)}{n!}$$

for each $n \geq 0$.

Proof. Evaluate the power series for $f^{(n)}$ at $x = c$. Every term is 0, except the first one, which is

$$f^{(n)}(c) = \frac{n!}{0!} a_n.$$

Since $0! = 1$, we obtain the desired equality. $\qquad\square$

7.4.6 Taylor Series

We now turn to the question of representing a given function by a power series. The following definition is suggested by Corollary 7.8 and also by the Taylor polynomial (5.5).

Definition 7.8. Suppose f is infinitely differentiable on an open interval I that contains a point c. The **Taylor series** of f at c is defined as the power series

$$T(f;c)(x) = \sum_{k=0}^{\infty} \frac{f^{(k)}(c)}{k!} (x - c)^k$$

for x in I.

There is no general claim that the Taylor series converges, or that it converges to the intended function f. An optimistic guess would be that the Taylor series converges to f in its interval of convergence. However, this is not true in general. In fact, it is possible for the Taylor series to converge to a function different from f. In other words, convergence of the Taylor series itself is not enough to imply that the limit is f. One such example is in the exercises.

Note that the degree n Taylor polynomial $T_n(x;c)$ (5.5) is the nth partial sum of the Taylor series of f. Thus, the convergence of the Taylor series is really about the convergence of the sequence of Taylor polynomials. Recall from (5.4) that the nth error term is defined as the difference

$$R_n(x;c) = f(x) - T_n(x;c).$$

Therefore, the Taylor series converges to f precisely when the sequence of error terms converges to 0. The following result gives a convenient sufficient condition for the Taylor series to converge to the desired function.

Theorem 7.11. *Suppose f is infinitely differentiable on an open interval I that contains a point c. Suppose for each x in I, there exists a real number M_x such that*

$$|f^{(n)}(y)| \le M_x \quad \text{for all } n \text{ and all } y \text{ between } c \text{ and } x.$$

Then the Taylor series $T(f;c)(x)$ converges pointwise to $f(x)$ on I.

Proof. For each x in I, by Taylor's Theorem 5.8 the nth error term is

$$R_n(x;c) = \frac{f^{(n+1)}(b)}{(n+1)!} (x - c)^{n+1}$$

for some point b between c and x. The assumption then implies

$$|R_n(x;c)| \le \frac{M_x|x-c|^{n+1}}{(n+1)!}.$$

This last sequence converges to 0 for each x, so $\lim_{n\to\infty} R_n(x;c) = 0$. This means that the Taylor series converges to f. □

For example, the previous theorem can be applied to the familiar functions e^x, $\sin x$, and $\cos x$ to show that each of their Taylor series converges to the function itself on $\mathbb{R}$.

7.4.7 Exercises

(1) Prove the cases $L = 0$ and $L = \infty$ of Theorem 7.6.
(2) Prove the cases $R = 0$ and $R = +\infty$ of Theorem 7.8.
(3) Explain why the proof of Theorem 7.8 cannot be used to show that the power series converges uniformly on its interval of convergence.
(4) Write down all the details of the proof of Theorem 7.9.
(5) Write down all the details of the proof of Theorem 7.10.
(6) Write down all the details of the proof of Corollary 7.7.
(7) Suppose the power series $\sum a_i(x-c)^i$ and $\sum b_i(x-c)^i$ both converge to a function f on some open interval containing c. Prove that $a_n = b_n$ for all n. This exercise shows that the coefficients a_n are uniquely determined by the power series in its interval of convergence.
(8) Consider the function

$$f(x) = \begin{cases} e^{-\frac{1}{x^2}} & \text{if } x \ne 0, \\ 0 & \text{if } x = 0. \end{cases}$$

(a) Prove that f is infinitely differentiable on $\mathbb{R}$.
(b) Prove that the Taylor series $T(f;0)$ of f at 0 is the 0 function. Conclude that the Taylor series does not converge to f on any open interval containing 0.

(9) Derive the Taylor series at 0 of the functions e^x, $\sin x$, and $\cos x$. Then show that in each case the Taylor series converges to the function itself on $\mathbb{R}$.
(10) For each integer $n \ge 1$, derive the Taylor series at 0 of $\frac{1}{(1-x)^n}$ and determine its interval of convergence. Then prove that the Taylor series converges to the function itself on the interval of convergence.
(11) Repeat the previous exercise for the functions $\ln(1+x)$ and $\tan^{-1} x$.
(12) Suppose $f(x) = \sum a_i x^i$ has radius of convergence $R > 0$.

(a) Prove that f is an odd function if and only if $a_i = 0$ for all even i.
(b) Prove that f is an even function if and only if $a_i = 0$ for all odd i.

7.5   Additional Exercises

(1) Suppose $\{f_n\}$ converges uniformly to f on S and on T. Prove that $\{f_n\}$ converges uniformly to f on $S \cup T$.

(2) Suppose $\sum f_n$ converges uniformly to f on S and on T. Prove that $\sum f_n$ converges uniformly to f on $S \cup T$.

(3) Suppose $\sum f_n$ converges uniformly to f on S with each f_n bounded. Prove that f is bounded on S.

(4) Suppose $\{f_n\}$ converges uniformly to f on S, and each f_n is bounded on S. Define $g_n = \frac{1}{n} \sum_{i=1}^{n} f_i$. Prove that $\{g_n\}$ converges uniformly to f on S.

(5) Suppose $\{f_n\}$ converges to f uniformly on $[a, b]$, each f_n is integrable on $[a, b]$, and $\{x_n\}$ is an increasing sequence in $[a, b]$ that converges to b. Prove that

$$\lim_{n \to \infty} \int_a^{x_n} f_n = \int_a^b f.$$

(6) Give an example in which f_n is not continuous at any point on $[0, 1]$, and $\sum f_n$ converges uniformly to a continuous function on $[0, 1]$

(7) Suppose $\sum f_n$ converges uniformly on S to a function f. Prove that

$$\lim_{n \to \infty} \left(\sup\{|f_n(x)| : x \in S\} \right) = 0.$$

(8) A sequence of functions $\{f_n\}$ on S is **equicontinuous** if, for every $\epsilon > 0$, there exists $\delta > 0$ such that

$$|x - y| < \delta \text{ with } x, y \in S \quad \text{implies} \quad |f_n(x) - f_n(y)| < \epsilon \quad \text{for all} \quad n.$$

   (a) Suppose $\{f_n\}$ is an equicontinuous sequence of functions on S that converges pointwise to f. Prove that f is uniformly continuous on S.

   (b) Suppose $\{f_n\}$ is a sequence of continuous functions on $[a, b]$ that converges uniformly to f. Prove that $\{f_n\}$ is equicontinuous.

(9) **Abel's Theorem** says: Suppose $f(x) = \sum a_i x^i$ has radius of convergence $R = 1$ and that it converges at $x = 1$. Then it converges uniformly on $[0, 1]$. Prove this theorem as follows.

   (a) For non-negative integers $m \le n$, use the abbreviation

$$a_{m,n} = a_m + a_{m+1} + \cdots + a_n.$$

   Prove that

$$\sum_{i=n}^{n+j} a_i x^i = a_{n,n+j} x^{n+j} + x^n (1 - x) \sum_{i=0}^{j-1} a_{n,n+i} x^i$$

   for all n and j. This is called **Abel's Formula**.

   (b) Use the convergence of $\sum a_i$ and the previous part to show that the power series $\sum a_i x^i$ satisfies the Cauchy criterion (Corollary 7.1) on $[0, 1]$.

(10) Use the previous exercise and a change of variable to prove the general form of Abel's Theorem: Suppose $f(x) = \sum a_i x^i$ has radius of convergence $0 < R < \infty$ and that it converges at $x = R$. Then it converges uniformly on $[0, R]$. Moreover, if it converges at $x = -R$, then it converges uniformly on $[-R, 0]$.

(11) Use the previous exercise to prove the statement: Suppose $f(x) = \sum a_i x^i$ has radius of convergence $0 < R < \infty$ and that it converges at $x = R$. Then f is continuous at $x = R$. If it converges at $x = -R$, then f is continuous at $x = -R$.

(12) Use the previous exercise and the Taylor series at 0 of $\ln(1+x)$ to prove the equality

$$\ln 2 = \sum_{n=1}^{\infty} (-1)^{n+1}\frac{1}{n} = 1 - \frac{1}{2} + \frac{1}{3} - \frac{1}{4} + \cdots.$$

In other words, the alternating harmonic series converges to $\ln 2$.

(13) Suppose $\{f_n\}$ is a decreasing sequence of continuous functions on $[a,b]$ (i.e., $f_n(x) \geq f_{n+1}(x)$ for all n and x) that converges pointwise on $[a,b]$ to the 0 function. Prove that f_n converges to 0 uniformly.

(14) Suppose $\{f_n\}$ is an increasing sequence of continuous functions on $[a,b]$ (i.e., $f_n(x) \leq f_{n+1}(x)$ for all n and x) that converges pointwise to a continuous function f on $[a,b]$. Prove that f_n converges uniformly to f. This result is known as **Dini's Theorem**.

(15) Give an example to show that the conclusion of the previous exercise may fail if the interval is of the form $[a,b)$.

(16) Give an example to show that the increasing assumption cannot be omitted in Dini's Theorem.

Hints for Selected Exercises

Section 1.1.4

(1) No. (10) 2^n.

Section 1.2.5

(1) No. (6) m^n and $\binom{m}{n}n!$.

Section 1.3.7

(20) Pick N with $x + N > 0$, and apply the $x > 0$ case to $x + N < y + N$.

Section 1.5.3

(6) Prove this by induction.

(14) Try to imitate the proof of Theorem 1.9.

(17) Prove this by contradiction.

Section 1.6

(7) First prove that S has a countable subset $A = \{a_0, a_1, \ldots\}$ (Exercise (6) on page 6). Define $T = S \smallsetminus \{a_0\}$. Let $f: S \to T$ be the function defined as

$$f(x) = \begin{cases} x & \text{if } x \notin A, \\ a_{n+1} & \text{if } x = a_n \in A. \end{cases}$$

Prove that f is a bijection.

(25) A is bounded above by b_1, and B is bounded below by a_1. To prove $\alpha \leq \beta$, first prove that $a_i < b_j$ for all i and j.

Section 2.1.5

(13) Use any $\epsilon > 0$ with $\epsilon < L$.

Section 2.2.4

(13) For the first part, note that $\binom{n}{2} = \frac{n(n-1)}{2}$ and that $b_n > 0$ for $n > 1$. To prove $b_n \to 0$, first establish

$$\frac{2}{n-1} > b_n^2.$$

(15) For the first part, first prove $(1 + r)^n > \binom{n}{2}r^2$.

Section 2.3.4

(8) First construct a subsequence that converges to 0.

(18) To show that $\{b_n\}$ is decreasing, prove that if $S \subseteq T$ are non-empty and bounded above, then $\sup(S) \leq \sup(T)$.

Section 2.4.5

(11) Use induction to construct a subsequence with $|a_{n_k} - l_k| < \frac{1}{k}$.

(16) Since $\lim \sup a_n = \lim s_n < M$, there exists N such that $n \geq N$ implies $s_n < M$.

Section 2.5

(17) Use an enumeration of the rational numbers between 0 and 1.

Section 3.1.5

(13) Use the Monotone Convergence Theorem.

(15) The sequence of partial sums for $\sum b_n$ is a subsequence of the sequence of partial sums for $\sum a_n$.

Section 3.2.3

(4) Use Cauchy's criterion and the Triangle Inequality.

Section 3.3.1

(2) and (3) Both statements are false.

Section 3.4.4

(6) For the first part, use the equality

$$\frac{a_{N+1}}{a_N} \frac{a_{N+2}}{a_{N+1}} \cdots \frac{a_n}{a_{n-1}} = \frac{a_n}{a_N}.$$

For the second part, show that $\lim \sup a_n^{\frac{1}{n}} \leq L + \epsilon$.

Section 3.5.1

(2) Prove the first part by contradiction. If there are only finitely many q_n, then after excluding finitely many a_k we have $\sum a_n = \sum p_n$, which is convergent, and hence absolutely convergent. For the second part, note that $\{p_n\}$ is a subsequence of $\{a_n\}$. The last part is proved by contradiction again.

Section 3.6

(1) Use Cauchy Condensation Test.

(11) For the last part, use the previous part and the Comparison Test.

(12) For the second part, use Kummer's Test.

Section 4.1.1

(5) Prove the "only if" part by proving its contrapositive.

Section 4.2.5

(10) If $\lim_{x \to c} f = L > 0$, use $\epsilon = \frac{L}{2}$.

Section 4.3.3

(3) Use the corresponding statements about limits of sequences.

Section 4.4.1

(6) Consider the function $g(x) = f(x) - x$, and use the Intermediate Value Theorem.

Section 4.5.1

(7) For the first part, use the previous exercise. For the second part, consider the sequence $x_1, y_1, x_2, y_2, \ldots$.

Section 4.6.4

(3) For the first part, if f is not strictly increasing, then there are $x < y$ in

$[a, b]$ such that $f(x) \geq f(y)$. So either $x \neq a$ or $y \neq b$. In the former case, we have $a < x < y \leq b$. Now apply the Intermediate Value Theorem on $[a, x]$ for the value $f(y)$ to obtain a contradiction.

Section 4.7.4

(12) Use induction on the number of points added to P.

Section 4.8

(10) Use the Intermediate Value Theorem on the function

$$g(x) = f(x + 1) - f(x)$$

on $[0, 1]$.

(18) For the first part, pick a sequence of irrational numbers that converges to x. For the second part, first show that every ϵ-neighborhood of x contains only finitely many rational numbers $\frac{p}{q}$ such that $q \leq \frac{1}{\epsilon}$. Pick δ such that the δ-neighborhood of x contains no such rational numbers.

Section 5.1.6

(2) The sequential definition of $f'(c)$ is this: For every sequence $\{x_n\}$ in $I \setminus \{c\}$ with $\lim x_n = c$, the sequence whose nth term is

$$\frac{f(x_n) - f(c)}{x_n - c}$$

converges to $f'(c)$. The ϵ-δ definition of $f'(c)$ is this: For every $\epsilon > 0$, there exists $\delta > 0$ such that $0 < |x - c| < \delta$ with x in I implies

$$\left| \frac{f(x) - f(c)}{x - c} - f'(c) \right| < \epsilon.$$

(8) Use induction and the product rule. (13) No.

Section 5.2.4

(2) Consider the function $(1 + x)^n$ on $[0, a]$.

(12) Apply the Mean Value Theorem on $[a, c]$ and $[c, b]$.

(14) and (15) First prove the case with $y = 1$. Then apply this case with x replaced by $\frac{x}{y}$.

(16) Use the continuity of f' to show the existence of J on which $f' \neq 0$. Then use Rolle's Theorem.

Section 5.3.5

(13) Use the previous exercise, and imitate the proof of L'Hospital's Rule, using the Mean Value Theorem $n + 1$ times.

Section 5.4

(7) For "only if" imitate the proof of chain rule. Define

$$g(x) = \begin{cases} \frac{f(x) - f(c)}{x - c} & \text{if } x \neq c, \\ f'(c) & \text{if } x = c. \end{cases}$$

(9) Apply Rolle's Theorem to $g(x) = e^{-rx} f(x)$.

(11) Set $g(x) = x$.

(13) Use Theorem 4.10 and the Intermediate Value Property for derivative.

(15) Use the Mean Value Theorem three times, twice for f and once for f'.

Section 6.1.6

(6) For the first inequality, use

$$f(x) + g(y) \le u_i(f) + u_i(g)$$

for all x, y in $[x_{i-1}, x_i]$.

(12) Choose P with norm $< \frac{\epsilon}{f(b)-f(a)}$.

(14) Prove this by induction on n.

(15) Use Theorem 6.2, the previous exercise, and an $\frac{\epsilon}{2}$-argument.

Section 6.2.3

(4) Use an $\frac{\epsilon}{2}$-argument.

(5) The "only if" part is an easy $\frac{\epsilon}{2}$-argument. The other direction is a slightly harder $\frac{\epsilon}{2}$-argument using Theorem 2.10.

(7) Use the approximation

$$\mathcal{R} \int_a^b f \approx R(f; P)$$

for P having small norm.

Section 6.3.4

(1) To prove that rf is integrable for $r > 0$, first prove

$$U(rf; P) = rU(f; P) \quad \text{and} \quad L(rf; P) = rL(f; P).$$

To prove that rf is integrable for $r < 0$, first prove

$$U(rf; P) = rL(f; P) \quad \text{and} \quad L(rf; P) = rU(f; P).$$

To prove $\int_a^b f \ge 0$ if $f(x) \ge 0$ for all x, first prove $U(f; P) \ge 0$.

(2) To prove that $|f|$ is integrable, first prove

$$u_i(|f|) - l_i(|f|) \le u_i(f) - l_i(f).$$

To prove the inequality, use $-|f| \le f \le |f|$ and Theorem 6.6.

(5) First prove the following statement. If $f = 0$ except for finitely many points on $[a, b]$, then f is integrable with $\int_a^b f = 0$.

(6) Use the Mean Value Theorem for integrals.

(8) Prove this by contradiction. If $f(x) > 0$ for some point x, then there is a small interval containing x on which $f \ge r$ for some positive real number r. Construct a partition P that includes the end points of this interval, and observe that $L(f; P) > 0$. Use this to conclude that $\int_a^b f > 0$.

(9) Note that g is uniformly continuous on $[c, d]$. We may assume that g is not the 0 function. Thus, given $\epsilon > 0$, there exists $0 < \delta < \epsilon_0$ such that

$$|x - y| < \delta \quad \text{implies} \quad |g(x) - g(y)| < \epsilon_0 = \frac{\epsilon}{2(b - a) + 4M},$$

where $M = \sup\{|g| \text{ on } [c, d]\} > 0$. Use the integrability of f to obtain a partition P such that

$$U(f; P) - L(f; P) < \delta^2.$$

Then prove that

$$U(g \circ f; P) - L(g \circ f; P) < \epsilon.$$

(10) Use the previous exercise with $g(x) = x^2$.

(11) First prove $fg = \frac{1}{4}[(f + g)^2 - (f - g)^2]$.

(12) Start with the inequality

$$mg(x) \le f(x)g(x) \le Mg(x),$$

where $m = \inf\{f(x) : x \in [a, b]\}$ and $M = \sup\{f(x) : x \in [a, b]\}$. Integrate and use the Intermediate Value Theorem for f.

Section 6.4.7

(1) Use Theorem 6.12 and take derivative on both sides to obtain $f(x) = -f(x)$.

(2) First write g in terms of $F(x) = \int_0^x f$.

Section 6.5

(1) First compute derivatives on both sides.

(3) A function of bounded variation is the difference of two increasing functions (Theorem 4.16), which are integrable by Exercise (12) on page 141.

(6) For the first part, use Exercise (7) on page 98. For the second part, use Exercise (5) on page 149 and Theorem 6.7.

(7) Use Exercise (13) on page 13, Exercise (5) on page 5, and Theorem 6.7.

(8) For the "if" part, take a large enough closed sub-interval of (a, b) and use an $\frac{\epsilon}{3}$-argument. $\int_a^b f = \lim_{\delta \to 0^+} \int_{a+\delta}^{b-\delta} f$

(9) Use Theorem 6.2, Theorem 6.4, and an $\frac{\epsilon}{3}$-argument.

(10) Use the step function in (4.5).

(12) Take f as the Thomae function on $[0, 1]$ (Exercise (14) on page 149). Define g as $g(0) = 0$ and $g(x) = 1$ for $x > 0$.

Section 7.1.6

(12) Use the approximations and inequality

$$f(x) \approx f_n(x) \le f_n(y) \approx f(y)$$

for $x < y$.

(13) Approximate $f(a)$ with $f_n(a)$ and then with $f_n(a_n)$.

Section 7.2.5

(3) Use an argument similar to the proof of the product rule.

(4) Enumerate the rational numbers in $[0, 1]$ as $q_1, q_2, \dots$. Define

$$f_n(x) = \begin{cases} 1 & \text{if } x = q_1, \dots, q_n, \\ 0 & \text{otherwise.} \end{cases}$$

Show that f_n is integrable, and that f is the characteristic function on $\mathbb{Q}$.

(12) Try $f_n = \frac{1}{n}\chi_{\mathbb{Q}}$.

Section 7.3.6

(3) Justify and use the equalities

$$\sum_{i=1}^{\infty} \left(\int_a^b f_i \right) = \lim_{n \to \infty} \sum_{i=1}^{n} \int_a^b f_i = \lim_{n \to \infty} \int_a^b \sum_{i=1}^{n} f_i = \int_a^b \lim_{n \to \infty} \sum_{i=1}^{n} f_i = \int_a^b f.$$

(8) Use Corollary 7.1 with $n = m + 1$.

Section 7.4.7

(3) The Weierstrass M-Test would not apply when $M_n = 1^n$.

(8) Prove by induction that

$$f^{(n)}(x) = \begin{cases} 0 & \text{if } x = 0, \\ e^{-1/x^2} x^{-3n} P_n(x) & \text{if } x \neq 0, \end{cases}$$

where P_n is some non-zero polynomial of degree $< 3n$.

Section 7.5

(5) Use an $\frac{\epsilon}{4}$-argument and the approximations

$$\int_a^b f \approx U(f; P) \approx U(f_n; P) \approx \int_a^b f_n \approx \int_a^{x_n} f_n.$$

(6) Try $f_n = \frac{1}{2^n} x \chi_{\mathbb{Q}}$.

(8) For the first part, use the approximations

$$f(x) \approx f_n(x) \approx f_n(y) \approx f(y).$$

For the second part, use the approximations

$$f_n(x) \approx f(x) \approx f(y) \approx f_n(y).$$

(9) For the first part, expand the right-hand side in terms of a_k. For the second part, first use the Cauchy criterion on the convergent series $\sum a_i$ to conclude $|a_{m,n}| < \epsilon$ for large m and any $n \geq m$.

(11) Reduce to the situation of the previous exercise by considering $g(x) = f(Rx)$.

List of Notations

Bibliography

Gelbaum, B. R. and Olmsted, J. M. H. (1964). *Counterexamples in Analysis* (Holden-Day, San Francisco).

Halmos, P. R. (1974). *Naive Set Theory* (Springer, New York).

Hobson, E. W. (1907). *The Theory of Functions of a Real Variable and the Theory of Fourier's Series* (Cambridge University Press, Cambridge).

Rudin, W. (1976). *Principles of Mathematical Analysis* (McGraw-Hill, New York).

Sprecher, D. A. (1987). *Elements of Real Analysis* (Dover, New York).

Index

Riemann Rearrangement Theorem, 79
Riemann sum, 142
Rolle's Theorem, 123
Root Test, 73

sequence, 30
  bounded, 37
  Cauchy, 47
  contractive, 50
  decreasing, 38
  Fibonacci, 30
  increasing, 38
  monotone, 38
  strictly decreasing, 38
  strictly increasing, 38
  strictly monotone, 38
  subsequence, 45
sequence of functions, 157
series, 62
  p-series, 65
  absolutely convergent, 72
  alternating, 70
  alternating harmonic, 70
  conditionally convergent, 72
  geometric, 30, 64
  harmonic, 64
  rearrangement, 77
series of functions, 169
set, 1
  closed, 26, 113
  countable, 22
  difference, 3
  disjoint, 3
  disjoint union, 3
  empty, 1
  finite, 21
  infinite, 21
  intersection, 3

non-empty, 1
open, 26, 113
power, 4, 26
product, 3
proper subset, 2
subset, 2
symmetric difference, 25
uncountable, 22
union, 2
Squeeze Theorem, 42
subsequence, 45
  limit, 52
substitution, 154
supremum, 11

Taylor polynomial, 127
Taylor series, 179
Taylor's Theorem, 127
  integral form, 152
telescoping sum, 49
Tietze Extension Theorem, 114
Triangle Inequality, 9

unbounded, 10
uniform convergence, 159
uniform limit, 159
uniformly bounded, 169
uniformly Cauchy, 160
upper bound, 10
  least, 11
upper sum, 136
  with respect to P, 136

variation, 105

Weierstrass M-Test, 170
Well-Ordering Property, 17